MOTORCYCLES

Arai
1125R

MOTORCYCLES

Published by Sandcastle Books Ltd

Orchard House,
6, Butt Furlong
Fladbury
Worcestershire, UK
WR10 2QZ

Produced by TAJ Books International LLP

27, Ferndown Gardens,
Cobham,
Surrey,
UK,
KT11 2BH

www.tajbooks.com

The Publishers wish to thank all of the manufacturers covered in this book for their cooperation in supplying photography.

All notations of errors or omissions (author inquiries, permissions) concerning the content of this book should be addressed to TAJ Books 27, Ferndown Gardens, Cobham, Surrey, UK, KT11 2BH, info@tajbooks.com.

ISBN-13: 978-1-906536-17-6

Printed in China.

CONTENTS

INTRODUCTION

Sunbeam

The modern motorcycle is descended from the pedal cycle. Its origins are lost in the mists of time, but it is generally acknowledged to have been invented by the German Karl von Drais in about 1817. This was a very basic machine propelled by pushing it along with ones feet.

In 1840 a Scottish blacksmith Kirkpatrick MacMillan made a machine with pedals and cranks. Subsequently Pierre Michaux in France designed the "Velocipede" which had pedals attached to the front wheel.

John Kemp Starley an Englishman manufactured in 1885 what is regarded as the modern bicycle with gears and a chain drive and brakes of course.

It is from this that the motorcycle as we know it today derived. Early machines were basically bicycles with a small motor inserted in the frame.

The first of these was made by Gottlieb Daimler and Wilhelm Maybach in Bad Cannstatt, Stuttgart in 1885. They named it the Reitwagen (Riding Car).

However an American, Sylvester Howard Roper had demonstrated a steam driven two wheel vehicle at fairs and circuses in 1867.

Neither of the above were sold to the public. The first put on sale was made by Hildebrand and Wolfmuller in 1894. The other major development was the invention of the pneumatic tyre by John Boyd Dunlop in 1888.

All the ingredients were now available leading to the birth of the mass-produced motorcycle, frame, seat, pneumatic tyres and a method of transmission of mechanical propulsion.

EarlyDays

Up to the beginning of the first world war Indian in the United States was the major producer of motor cycles, with production peaking at 20,000 a year. By 1920 Harley-Davidson had assumed the lead with dealers in 67 countries.

In 1928 the German maker DKW had taken over as the largest manufacturer.

After the second world war the British BSA Group had the lead producing 75,000 machines a year until 1955.

The German NSU Motorenwerke was the worlds major manufacturer until the 1970s.

The Japanese makers Honda, Suzuki, Yamaha and Kawasaki then assumed the mantle of mass manufacture, although Harley-Davidson, Triumph, MotoGuzzi still maintain a major presence in the niche market.

The Mechanics

The technical specification of the motorcycle has not changed in its 100 year history.

A metal frame made from steel tube supports front and rear forks, these were originally unsprung, but eventually the front forks were independently sprung with friction dampers which were adjustable. This led to telescopic forks with internal oil dampers. The rear independent suspension followed soon after.

The brakes were of the drum variety similar to those on motor cars, but because of the restrictions on the size of these they were not as effective as one would expect for the speeds attained by the machine. The calliper disc brake became universal after its introduction on cars.

Engines on the early machines were usually of a small capacity and single cylinder, with notable exceptions by Indian and Harley-Davidson who had twin-cylinder V configuration and Ariel who had a square four layout.

These engines were side-valves operated by push-rods open to the elements extremely noisy and distributing oil over the riders.

Most of these engines were air-cooled but some makers such as Scott had water-cooled installations, which were quieter and cleaner.

The engines had vertical cylinders, but Douglas developed the flat-twin very successfully, with BMW eventually designing a similar concept. The engines were principally made from cast-iron with crank-cases of aluminium. The pistons originally

1908 Triumph

Harley Davidson

made from iron changed to aluminium, while the valves were made from steel alloys. Lubrication was originally by vegetable derived oils but, became mineral based eventually. Ignition was by magneto to a spark plug with eventually a circuit breaker and coil was introduced. Fuel was by a carburetor with a float in a chamber under a gravity-fed tank, jets then introduced the fuel to the engine by an inlet manifold. Exhaust was into a pipe leading to a silencer exhausting at the rear of the machine. Two valves per cylinder was the norm.

Clutches were leather based initially but developed into cork and asbestos faced plates. This developed into the all-metal clutch which serves today. They were controlled by a hand-operated lever on the handlebars.

Gears were non-existent on the very early machines. Then a separate gearbox driven by an adjustable chain from the engine was fitted. The most notable of these was the "Burman" originally with three speeds. Early machines had hand change mechanisms, these eventually developed into the foot change we are familiar with today. The original transmission was by belt on pullies, this changed to gears driving a roller chain to the back wheel, the tension was adjusted by altering the wheel whose spindle was located in slots in the rear frame

The first machines were started by being pushed by the rider, kick-starts were then introduced, together with a hand operated device to decompress the engine to enable easy starting.

Speed indication was obtained by handle-bar mounted speedometer, which on more expensive machines was augmented by a revolution counter.

An ampere meter was also mounted to indicate battery state.

Prior to batteries being installed lighting was by acetylene lamps, each lamp had a reservoir containing carbide and water generating gas, this was fed through a jet which was ignited by the rider.

A method of warning of approach developed from the bulb operated horn to the electric type in use today.

Modern Developments

Post second world war saw the major developments that are recognisable in today's machines.

Engines have become highly sophisticated, being multi-cylindered with overhead camshafts a common trend. Synthetic oils have been developed to cope with increases in temperatures and loads within the engine.

Gearboxes are now integral with the engine, sharing the same lubricants. Four and five speed ratios are common. Transmissions are by technically advanced roller chain with each roller sealed containing a lubricant. Shaft drive introduced by BMW is more common, also used by MotoGuzzi.

Wheels which were for many years spoked, are now more commonly die-cast aluminium. Brakes are disc with hydraulic cylinders and extremely efficient.

Frames are stiffer, with suspension front and rear having adjustable hydraulic damping.

Lighting has been improved beyond all recognition particularly with the introduction of quartz halogen bulbs.

Tyres are designed to deal with wet conditions, enabling higher speeds without mishap.

Many machines have fairings to protect the rider from the worst of the weather, but also leading o improved fuel economy.

Modern Manufacturers

Harley-Davidson can probably claim to be the longest US manufacturer in continual production. Triumph have been around for about eighty years but not in continuous production. The Japanese makers have been around longer than most people realise, Kawasaki being part of a large engineering conglomerate, Yamaha having its origins in musical instrument manufacture.

BMW built aircraft engines and cars before moving into the motorcycle field.

Moto Guzzi were made before the war and had a very good

The Bat 1,000 cc

INTRODUCTION

AJS

reputation in road racing after it. Husqvarna in Sweden have been made since 1903 so may claim to have been around longer than Harley-Davidson.

AJS 1910-1974

AJS was the name used for cars and motorcycles made by the Wolverhampton, England company A. J. Stevens & Co. Ltd, from 1909 to 1931, by then holding 117 motorcycle world records, and after the firm was sold the name continued to be used by Matchless, Associated Motorcycles and Norton-Villiers on four-stroke motorcycles till 1969, and since the names resale in 1974, on small capacity two-strokes.

Brough Superior 1920-1940

Brough Superior motorcycles and motor cars were made by George Brough in his Brough Superior works on Haydn Road in Nottingham, England from 1919 to 1940. They were dubbed the "Rolls-Royce of Motorcycles" by H. D. Teague of The Motorcycle newspaper. Approximately 3048 of 19 models were made in 21 years of production. In 2004, around 1000 still exist. T.E. Lawrence ("Lawrence of Arabia") owned seven bikes and died from injuries sustained while crashing one. George Bernard Shaw was another among many celebrities that were enthusiastic about Brough products.

BSA 1910-1973

BSA was founded in 1861 in the Gun Quarter, Birmingham, England by fourteen gunsmiths of the Birmingham Small Arms Trade Association, who had together supplied arms to the British government during the Crimean War. The company branched out as the gun trade declined; in the 1870s they manufactured the Otto Dicycle, in the 1880s the company began to manufacture bicycles and in 1903 the company's first experimental motorcycle was constructed. Their first prototype automobile was produced in 1907 and the next year the company sold 150 automobiles. By 1909 they were offering a number of motorcycles for sale and in 1910 BSA purchased the British Daimler Company for its automobile engines.

The first wholly BSA motorcycles were built in 1910, before then engines had come from other manufacturers. BSA Motorcycles Ltd was set up as a subsidiary in 1919. Initially, after World War II, BSA motorcycles were not generally seen as racing machines, compared to the likes of Norton. . In the immediate post war period few were entered in races such as the TT races, though this changed dramatically in the Junior Clubman event (smaller engine motorcycles racing over some 3 or 4 laps around one of the Isle of Man courses). In 1947 there were but a couple of BSA mounted riders, but by 1952 BSA were in the majority and in 1956 the makeup was 53 BSA, 1 Norton and 1 Velocette.

To improve US sales, in 1954, for example, BSA entered a team of riders in the 200 mile Daytona beach race with a mixture of single cylinder Gold Stars and twin cylinder Shooting Stars assembled by Roland Pike. The BSA team riders amazingly took first, second, third, fourth, and fifth places with two more riders finishing at 8th and 16th. This was the first case of a one brand sweep.

FN

FN (Fabrique Nationale de Herstal) was a Belgian company established in 1899 to make arms and ammunition, and from 1901 to 1967 was also a motorcycle manufacturer. FN manufactured the world's first four cylinder motorcycle, was famous for the use of shaft drive in all models from 1903 to 1923, achieved success in sprint and long distance motorcycle racing, and after 1945, also in motocross.

In 1899 FN made shaft and chain driven bicycles, and in 1900 experimented with a clip-on engine. In December 1901 the first 133 cc single cylinder motorcycle was built, followed in 1903 by a shaft driven 188 cc single cylinder motorcycle. In 1904 a 300 cc

1915 Indian

1914 Royal Enfield

single cylinder motorcycle was produced. In 1909 the two speed singles had camshafts to open the inlets, instead of the earlier "automatic" valves. Starting from 1912 the singles had a hand lever clutch and foot pedal rear brake.

The FN Four

In 1905 the first 362 cc shaft drive in-line FN inlet-over-exhaust four cylinder motorcycle appeared, designed by Paul Kelecom. This was the world's first manufactured four cylinder motorcycle. By 1907 the Four engine had grown to 412 cc, and that year's single cylinder 244 cc FN motorcycle was the first bike with a multiple ratio belt drive system, using a patented variable size engine pulley. For 1908, the US Export model began manufacture. The Four had a 493 cc engine, and in 1910 that became 498 cc. This bike weighed 75 kg (165 lb) dry, and could do 40 mph (64 km/h). The 1913 Fours had a two speed gearbox and clutch, at the rear of the shaft drive, and bicycle pedals were permanently replaced with footrests from then on. For 1914 the FN "Type 700" 748 cc Four was released, with the gearbox at the rear of the engine.

Husqvarna

As with many motorcycle manufacturers, Husqvarna first began producing bicycles in the late 19th century. In 1903, they made the jump to motorcycle manufacturing. In 1920 Husqvarna established its own engine factory and the first engine to be designed was a 550 cc four-stroke 50-degree side-valve V-twin engine, similar to those made by companies like Harley-Davidson and Indian. Although they once made motorcycles for street use, and raced at road circuits such as the Isle of Man TT prior to World War II, they are more well known for producing world championship winning motocross and enduro bikes. In the 1960s, their lightweight, two-stroke engined off-road bikes helped make the once dominant British four-stroke motorcycles obsolete. Throughout the 1960s and 1970s they were a dominant force in the motocross world, winning 14 Motocross world championships in the 125cc, 250cc and 500cc divisions and 24 enduro world championships.

The Husqvarna motorcycle division was sold to Italian motorcycle manufacturer Cagiva in 1987 and became part of MV Agusta Motor S.p.A. The motorcycles (widely known as a "Husky") are now produced in Varese. Husqvarna produces a diverse range of motocross, enduro and supermoto machines using their own two-stroke or four-stroke engines, ranging in capacity from 125cc to 576cc. Racing continues to be important to Husqvarna, competing in world enduro and world supermoto championships. Gerald Delepine, riding a Husqvarna SMR660, became supermoto world champion in 2005.

In July 2007 Husqvarna was purchased by BMW for a reported 93 million euros. BMW Motorrad plans to continue operating Husqvarna Motorcycles as a separate enterprise. All development, sales and production activities, as well as the current workforce, will remain in place at its present location at Varese.

Royal Enfield 1899-1970 (UK)

Royal Enfield was the brand of the Enfield Cycle Company, an English engineering company. Most famous for producing motorcycles, they also produced, bicycles, lawnmowers, stationary engines, and even rifle parts for the Royal Small Arms Factory in Enfield. This legacy of weapons manufacture is reflected in the logo, a cannon, and their motto "Made like a gun, goes like a bullet". It also enabled the use of the brand name Royal Enfield from 1890.

In 1955 Enfield of India started assembling Bullet motorcycles under licence from UK components, and by 1962 were manufacturing complete bikes. The original Redditch, Worcestershire - based company dissolved in 1970, but Enfield of India, based in Chennai, continued, and bought the rights to the Royal Enfield name in 1995. Royal Enfield production continues,

Zenith

INTRODUCTION

1910 Rex

and now Royal Enfield is considered as the oldest motorcycle company in the world still in production.

Sunbeam 1912-1957

Sunbeam was a British motorcycle marque generally known for high quality

John Marston, the man who started it all was born in Ludlow, Shropshire, U.K. in 1836, of a minor landowning family. In 1851 at age 15, he was sent to Wolverhampton to be apprenticed to Edward Perry as a japanware manufacturer. At the age of 23 he left and set up his own japanning business, John Marston Ltd, making any and every sort of domestic article. He did so well that when Perry died in 1871, Marston took over his company and incorporated it in his own.

The company began making bicycles, and on the suggestion of his wife Ellen, Marston adopted the trademark brand "Sunbeam". Consequently, the Paul Street works were called Sunbeamland. John Marston was a perfectionist, and this was reflected in the high build quality of the Sunbeam bicycle, which had an enclosure around the chain in which an oil bath kept the chain lubricated and clean. They were made until 1936, and to the end remained the best bicycle money could buy.

From 1903 John Marston Ltd had made some early experiments in adding engines to bicycles but they were unsuccessful, one man being killed. John Marston's aversion to motorcycles did not encourage further development, and so the Sunbeam Motor Car Company Ltd was founded in 1905. However, suffering from a slump which hit car making, Marston was pushed into making motorcycles from 1912 onwards (at the age of 76), for which there was a greater and increasing market. Following in the tradition of their bicycles, the motorcycles were of high-quality, usually with a single cylinder, and known as the "Gentleman's Machine." Sunbeam motorcycles performed well in the early days of the famous TT (Tourist Trophy) races in the Isle of Man.

After the First World War, the Marston company was sold to a consortium. In 1919 the consortium became part of Nobel Industries Limited. In 1927 Nobel Industries amalgamated with Brunner Mond Ltd. to form Imperial Chemical Industries (ICI). In this huge organization motorcycles were a small part.

In 1937 the Sunbeam motorcycle trademark was sold to Associated Motor Cycles Ltd ("AMC"), which continued to make Sunbeam bicycles and motorcycles until 1939. Other brandnames of motorcycles owned by AMC were Matchless, AJS, Norton, James, and Francis-Barnett.

In 1943, AMC sold the Sunbeam name to BSA, and Sunbeam Cycles Ltd came into being. Three Sunbeam motorcycle models were produced from 1946 to 1956, not in the main BSA factory at Small Heath, Birmingham, but in Redditch, Worcestershire. These were followed by two scooter models from 1959-1964. The new Sunbeam motorcycles were of an entirely new design inspired by BMW German army motorcycles captured in World War II.

Velocette 1909-1971

Velocette is the name given to motorcycles that were made by Veloce Ltd, in Hall Green, Birmingham, England.

One of several motorcycle manufacturers in Birmingham, Velocette was a small, family-owned firm, selling far fewer hand-built motorcycles than the giant BSA, Norton, or Triumph concerns. Renowned for the quality of their products, the company was 'always in the picture' in international motorcycle racing, from the mid-1920s through the 1950s, culminating in two world championship titles (1949–1950 350 cc) and their legendary and still-unbeaten 24 hours at 100 mph (161 km/h) record. Veloce, while small, was a great technical innovator and many of their patented designs are commonplace on motorcycles today, including the positive-stop foot shift and swingarm rear fork with hydraulic shocks

The company was founded by John Taylor (born Johannes Gütgemann and later known as John Goodman), and William

Velocette

Douglas

Gue as "Taylor, Gue Ltd." in 1905. Their first motorcycle was the Veloce. Later that year, John Taylor set up Veloce Limited, to produce cycles and related products and services. Veloce Ltd initially produced four-stroke motorcycles. The first two-stroke, built in 1913, was called a Velocette. This name was used for all of their subsequent models.

Vincent-HRD 1928-1955

HRD was founded by the British (RFC) pilot, Howard Raymond Davies, who was shot down and captured by the Germans in 1917. Legend has it that it was while a prisoner of war that he conceived the idea of building his own motorcycle, and contemplated how he might achieve that. It was not until 1924 that Davies entered into partnership with E J Massey, trading as HRD Motors. Various models were produced, generally powered by JAP (J A Prestwich) engines.

Unfortunately, even though HRD motorcycles won races the company ran at a loss, and in January 1928 it went into voluntary liquidation. The company was initially bought by Ernest Humphries of OK-Supreme Motors for the factory space, and the HRD name, jigs, tools, patterns, and remaining components were subsequently offered for sale again.

Indian 1901-1953

The Indian Motocycle Manufacturing Company (sic) was a motorcycle manufacturer in Springfield, Massachusetts. Indian was America's oldest motorcycle brand and was once the largest manufacturer of motorcycles in the world. The most popular models were the Scout, made prior to WWII, and the Chief, which had its heyday from 1922-53

Harley-Davidson 1903-

Harley-Davidson Motor Company is an American manufacturer of motorcycles based in Milwaukee, Wisconsin. The company sells heavyweight (over 750 cc) motorcycles designed for cruising on the highway. Harley-Davidson motorcycles (popularly known as "Harleys") have a distinctive design and exhaust note. They are especially noted for the tradition of heavy customization that gave rise to the chopper-style of motorcycle.

Harley-Davidson attracts a loyal brand community, with licensing of the Harley-Davidson logo accounting for almost 5% of the company's net revenue ($41 million in 2004). In 2003, the Buell Motorcycle Company became a wholly-owned subsidiary of Harley-Davidson, the same year that the Motor Company celebrated its 100th birthday. The Motor Company supplies many American police forces with their motorcycle fleets.

MotoGuzzi 1921-

Moto Guzzi (aka "Guzzi") is an Italian motorcycle manufacturer that has endured from the industry's infancy to its place today as the oldest European manufacturer in continuous motorcycle production. Guzzi is now one of seven brands owned by Piaggio & Co. SpA, Europe's largest motorcycle manufacturer and the world's fourth largest motorcycle manufacturer by unit sales.

Established in 1921 in Mandello del Lario, Italy, Moto Guzzi has led Italy's motorcycling manufacture, enjoyed prominence in world-wide motorcycle racing, and led the industry in ground-breaking innovation — for the greater part of its history.

Today Moto Guzzi impresses its heritage on a range of motorcycles in touring, cruising, racing and naked configurations — each with the company's iconic, air-cooled 90° V-twin engines.

Gilera 1909-

Gilera is an Italian motorcycle manufacturer founded in Arcore in 1909 by Giuseppe Gilera. In 1969 the company was purchased by the Piaggio & Co. SpA -- which now holds six marques and is the world's fourth largest motorcycle manufacturer.

In 1935 Gilera acquired rights to the Rondine four-cylinder engine. This formed the basis for Gileras racing machines for nearly forty years. From the mid-thirties Gilera developed a range of four-stroke engine machines. The engines ranged from 100 to 500cc. The most famous of which was the 1939 Saturno.

After World War II, Gilera dominated Grand Prix motorcycle racing, winning the 500cc road racing world championship 6 times in 8 years. Facing a downturn in motorcycle sales due to the increase in the popularity of automobiles after the war, Gilera made a gentleman's agreement with the other Italian motorcycle makers to quit Grand Prix racing after the 1957 season as a cost cutting measure.

In 1992, Gilera made a return to the Grand Prix arena and Piaggio continues to produce small-displacement motorcycles with the Gilera name.

INTRODUCTION

Honda 1938-

Honda's first motorcycle to be put on sale was the 1947 A-Type (one year before the company was officially founded). However, Honda's first full-fledged motorcycle on the market was the 1949 Dream D-Type. It was equipped with a 98cc engine producing around 3 horsepower (2.2 kW). This was followed by other highly popular scooters throughout the 1950s.

In 1958, the American Honda Company was founded and one year later, Honda introduced its first model in the United States, the 1959 Honda C100 Super Cub. The Honda Cub holds the title of being the best-selling vehicle in history, with around 50 million units sold around the world.

Kawasaki 1953-

Kawasaki Heavy Industries Consumer Products and Machinery Company is the Consumer Products and Machinery production division of Kawasaki Heavy Industries. It produces Motorcycles, ATVs, Utility vehicles, Jet Ski personal watercrafts, General-purpose gasoline engines.

Kawasaki's Aircraft Company began the development of a motorcycle engine in 1949. The development was completed in 1952 and mass production started in 1953. The engine was an air-cooled, 148cc, OHV, 4-stroke single cylinder with a maximum power of 4 PS (3.9 hp/2.9 kW) at 4,000 rpm.

In 1954 the first complete Kawasaki Motorcycle was produced under the name of Meihatsu, a subsidiary of Kawasaki Aircraft.

In 1960 Kawasaki completed construction of a factory dedicated exclusively to motorcycle production and bought Meguro Motorcycles.

Kawasaki has since then become one of the world's major motorcycle manufacturers.

Suzuki 1909-

In 1909, Michio Suzuki founded the Suzuki Loom Company in the small seacoast village of Hamamatsu, Japan. Business boomed as Suzuki built weaving looms for Japan's giant silk industry. Suzuki's only desire was to build better, more user-friendly looms. In 1929, Michio Suzuki invented a new type of weaving machine, which was exported overseas. Suzuki filed as many as 120 patents and utility model rights. For the first 30 years of the company's existence, its focus was on the development and production of these exceptionally complex machines.

But the joy was short-lived as the cotton market collapsed in 1951.

Faced with this colossal challenge, Suzuki's thoughts went back to motor vehicles. After the war, the Japanese had a great need for affordable, reliable personal transportation. A number of firms began offering "clip-on" gas-powered engines that could be attached to the typical bicycle. Suzuki's first two-wheel ingenuity came in the form of a motorized bicycle called, the "Power Free." Designed to be inexpensive and simple to build and maintain, the 1952 Power Free featured a 36 cc two-stroke engine. An unprecedented feature was the double-sprocket gear system, enabling the rider to either pedal with the engine assisting, pedal without engine assist, or simply disconnect the pedals and run on engine power alone. The system was so ingenious that the patent office of the new democratic government granted Suzuki a financial subsidy to continue research in motorcycle engineering, and so was born Suzuki Motor Corporation.

In 1953, Suzuki scored the first of countless racing victories when the tiny 60 cc "Diamond Free" won its class in the Mount Fuji Hill Climb.

By 1954, Suzuki was producing 6,000 motorcycles per month and had officially changed its name to Suzuki Motor Co., Ltd. Following the success of its first motorcycles, Suzuki created an even more successful automobile: the 1955 Suzulight. Suzuki showcased its penchant for innovation from the beginning. The Suzulight included front-wheel drive, four-wheel independent suspension and rack-and-pinion steering -- features common on cars half a century later.

Yamaha 1955-

Yamaha Motor Company Limited, a Japanese motorized vehicle-producing company (whose HQ is at 2500 Shingai, Iwata, Shizuoka), is part of the Yamaha Corporation. After expanding Yamaha Corporation into the world's biggest piano maker, then Yamaha CEO Genichi Kawakami took Yamaha into the field of motorized vehicles on July 1, 1955. The company's intensive research into metal alloys for use in acoustic pianos had given Yamaha wide knowledge of the making of lightweight, yet sturdy and reliable metal constructions. This knowledge was easily applied to the making of metal frames and motor parts for motorcycles. Yamaha Motor is the world's second largest producer of motorcycles. It also produces many other motorized vehicles such as all-terrain vehicles, boats, snowmobiles, outboard motors, and personal watercraft.

BMW 1923-

BMW Motorrad, a subsidiary of BMW, manufactures motorcycles. Originally an aircraft engine manufacturer at the turn of the last century and through World War I, BMW introduced the first motorcycle under its name, the R32, in 1923.

Although BMW motorcycles have been long associated with their original engine configuration, the flat-twin or boxer engine, the company today manufactures a full line of motorcycles in a variety of engine and riding configurations

BMW began as an aircraft engine manufacturer before World War I. With the Armistice, the Treaty of Versailles banned any German air force and thus need for aero engines, so the company turned first to making air brakes, agricultural machinery, toolboxes and office furniture. Dissatisfied with that, they eventually turned to manufacturing motorcycles. After the MB215 engine and the two stroke "Flink", 1923 saw the arrival of a complete motorcycle under the BMW name, the R 32.

Max Friz, BMW's chief designer, turned to motorcycle and car engines. Within four weeks, he had copied the now-legendary opposing flat twin cylinder engine which we know today as the boxer engine. This product was the second revolutionary product that Friz copied that firmly placed BMW AG in a profitable position.

The first boxer engine was the fore-and-aft M2B15, based on a British Douglas design. It was manufactured by BMW in 1921–1922 but mostly used in other brands of motorcycles, notably Victoria of Nuremberg. The M2B15 proved to be moderately successful and BMW used it in its own Helios motorcycle. BMW also developed and manufactured a small 2-stroke motorcycle called the Flink for a short time.

DKW 1919-1958

Dampf Kraft Wagen (German: steam-driven car) or DKW is a historic car and motorcycle marque. In 1916, the Danish engineer Jørgen Skafte Rasmussen founded a factory in Saxony, Germany, to produce steam fittings. In the same year, he attempted to produce a steam-driven car, called the DKW. Although unsuccessful, he made a two-stroke toy engine in 1919, called Des Knaben Wunsch — "a boy's desire". He also put a slightly modified version of this engine into a motorcycle and called it Das Kleine Wunder — "a little marvel". This was the real beginning of the DKW brand: by the 1930s, DKW was the world's largest motorcycle manufacturer.

In 1932, DKW merged with Audi, Horch and Wanderer to form the Auto Union. Auto Union came under Daimler-Benz ownership in 1957, and was finally purchased by the Volkswagen Group in 1964. The last DKW car was the F102 which ceased production in 1966; after this the brand was phased out.

NSU 1901-1958

NSU began as a knitting machine manufacturer in the town of Riedlingen on the Danube in 1873, and moved to Neckarsulm, where the river Sulm flows into the river Neckar, in 1884. The company soon began to produce bicycles as well, and by 1892, bicycle manufacturing had completely replaced the knitting machine production. At about this time, the name NSU (from Neckar and Sulm) appeared as brand name.

In the early years of the 20th century NSU motorcycles were developed, in 1905 the first NSU cars appeared. In 1932 the car production in Heilbronn was sold to Fiat.

During World War II NSU designed and produced the famous Kettenkrad, NSU HK101 a half-tracked motorcycle with the engine of the Opel Olympia.

After the war, NSU restarted in a completely destroyed plant with pre-war constructions like the Quick, OSL and Konsul motorbikes. And also still the HK101 could be purchased at NSU as an all terrain vehicle in a civil version. The first post war construction was the NSU Fox in 1949, available in a 2-stroke and a 4-stroke version. In 1953 the famous NSU Max followed, a 250 cc motorbike with a unique overhead camdrive with connecting rods. All these new models had a very innovative monocoque frame of pressed steel and a central rear suspension unit. Albert Roder, the genius chief engineer behind the success story, made it possible that in 1955 NSU became the biggest motorcycle producer in the world. NSU also holds 4 world records for speed: 1951, 1953, 1954 and 1955. In 1956 Wilhelm Herz started at the Bonneville Salt Flats, Utah. Herz was the first man to ride a motorcycle faster than 200 miles per hour, in August 1956. In 1957 NSU re-entered the car market with the new NSU Prinz, a small car with a doubled NSU Max engine, an air cooled two-cylinder engine of 600 cc and 20 hp. Motorbike production continued until 1968

The Sd. Kfz 2 Kleines Kettenkraftrad (Small-Tracked Motorcycle) was a light armored vehicle with the front end of a motorcycle attached to the rear end of a half-track. The handlebars operated differential brakes on the tracks as well as turning the front wheel. Powered by an 1500cc 4-cylinder Opel engine, it was produced by the thousands to equip Wehrmacht Panzer units.

Norton 1902-1999

Norton was a British motorcycle marque from Birmingham, founded in 1898 as a manufacturer of cycle chains.

By 1902 they had begun manufacturing motorcycles with bought-in engines. In 1908 a Norton built engine was added to the range. This began a long series of production of single cylinder motorcycles. They were one of the great names of the British motorcycle industry, producing machines which for decades dominated racing with highly tuned single cylinder engines under the Race Shop supremo Joe Craig.

Postwar a 500 cc twin cylinder model called the Dominator or Model 7 was added to the range for 1949, and this evolved into the 1970s through 500 cc, to 600 cc, to 650 cc, to 750 cc and to 850 cc models with the Dominator, 650, Atlas and Commando, all highly regarded road motorcycles of their time.

BMW

Martin Stolle with his Victoria powered by a M 2 B 15 engine ca. 1921

BMW began as an aircraft engine manufacturer before World War I. With the Armistice, the Treaty of Versailles banned any German air force and thus need for aero engines, so the company turned first to making air brakes, agricultural machinery, toolboxes and office furniture. Dissatisfied with that, they eventually turned to manufacturing motorcycles. After the MB215 engine and the two stroke "Flink", 1923 saw the arrival of a complete motorcycle under the BMW name, the R 32.

Max Friz, BMW's chief designer, turned to motorcycle and car engines. Within four weeks, he had copied the now-legendary opposing flat twin cylinder engine which we know today as the boxer engine. This product was the second revolutionary product that Friz copied that firmly placed BMW AG in a profitable position.

The first boxer engine was the fore-and-aft M2B15, based on a British Douglas design. It was manufactured by BMW in 1921–1922 but mostly used in other brands of motorcycles, notably Victoria of Nuremberg. The M2B15 proved to be moderately successful and BMW used it in its own Helios motorcycle. BMW also developed and manufactured a small 2-stroke motorcycle called the Flink for a short time.

With the development of its first light alloy cylinder head, a much more significant "across the frame" version of the boxer engine was designed. Using the new aluminium alloy cylinders, Friz designed the 1923 R 32 with a 486 cc engine with 8.5 hp (6.3 kW) and a top speed of 95–100 km/h (60 mph). The engine and gear box formed a bolt-up single unit. At a time when many motorcycle manufacturers used a total-loss oiling systems, the new BMW engine featured a recirculating wet-sump oiling system. However, it was not a "high-pressure oil" system based on shell bearings and tight clearances that we are familiar with, but a drip feed to roller bearings. This system was used by BMW until 1969. The wet-sump system was not overly common on motorcycles until the 1970s and the arrival of Japanese motorcycles. Until then, many manufacturers had used dry sump, with an external oil-tank made of sheet metal.

The R 32 became the foundation for all future boxer powered BMW motorcycles. BMW oriented the boxer engine with the cylinder heads projecting out on each side for cooling as per the earlier British ABC.

The R32 also incorporated a shaft drive. BMW continued to use shaft drives in all of its motorcycles until the introduction of the F 650 in 1994. The F 650 series, and later the F 800 series when introduced in 2006, featured either a chain or a belt drive system, both of which were a departure from BMW tradition.

BMW R2 Series 1 - 1930

By this time the benefits of overhead cams were known; higher revs could be obtained before the onset of valve float. However, the basic boxer design did not lend itself to overhead cams. To obtain the benefits of overhead cams without overly increasing the engine width, BMW incorporated a system that was so advanced for its racing bikes that it resurrected it many decades later in the R 1100 RS oilhead. The system was two cams in the head operating rocker arms via short push rods.

During World War II, the BMW motorcycle copies of the Zündapp KS750 performed exceptionally well in the harsh environment of the North African deserts. At the beginning of the war, the German army needed as many vehicles as it could get of all types. Although motorcycles of every style performed acceptably well in Europe, in the desert the protruding cylinders of the flat-twin engine and shaft drive performed better than vertical and V-twin engines, which overheated in the hot air, and chain-drives, which were damaged by desert sand.

Also during World War II, the U.S. Army asked Harley-Davidson, Indian, Delco, and Crossley to produce a motorcycle similar to BMW's side-valve R71. So Harley copied the BMW engine and transmission — simply converting metric measurements to inches — and produced the shaft-drive 750 cc 1942 Harley-Davidson XA.

The end of World War II found BMW in ruins. Its plant outside of Munich was destroyed by allied bombing. The Eisenach facility was not. It was not dismantled by the Soviets as reparations and sent back to the Soviet Union, to make IMZ-Ural motorcycles as is commonly alleged. The IMZ plant was supplied to the Soviets by BMW under licence prior to the commencement of the Great Patriotic War. After the war the terms of Germany's surrender forbade BMW from manufacturing motorcycles. Most of BMW's brightest engineers were taken to the US and Russia to continue their work on jet engines which BMW produced during the war.

When the ban on the production of motorcycles was lifted in Allied controlled Western Germany, BMW had to start from scratch. There were no plans, blueprints, or schematic drawings because they were all in Eisenach.. Company engineers had to use surviving pre-war motorcycles to create new plans. The first post-war BMW motorcycle in Western Germany, a 250 cc R 24, was produced in 1948. The R 24 was based on the pre-war R 23, and was the only postwar BMW with no rear suspension. In 1949, BMW produced 9,200 units and by 1950 production surpassed 17,000 units.

The situation was very different in Soviet-controlled Eastern Germany where the Eisenach plant was producing R35 and a handful of R 75 motorcycles for reparations. Eventually this plant became EMW (Eisenacher Motoren Werke).

In 1952, BMW introduced its first postwar sporting motorcycle, the R 68. Only 1,453 R 68s were made, making it the rarest postwar BMW motorcycle. It has a 594 cc single cam engine with 8.0:1 compression ratio and larger valves, together producing 35 hp (26 kW). The carburettor venturi throat sizes were 26 mm.

As the 1950s progressed, motorcycle sales plummeted. In 1957, three of BMW's major German competitors went out of business. In 1954, BMW produced 30,000 motorcycles. By 1957, that number was less than 5,500. However, by the late 50's, BMW exported 85% of its boxer twin powered motorcycles to the United States. At that time, Butler & Smith, Inc. was the

exclusive U.S. importer of BMW.

On June 8, 1959, John Penton rode a BMW R 69 from New York to Los Angeles in 53 hours and 11 minutes, setting a record. The previous record of 77 hours and 53 minutes was set by Earl Robinson on a 45 cubic inch (740 cc) Harley-Davidson.

Although U.S. sales of BMW motorcycles were strong, BMW was in financial trouble. Through the combination of selling off its aircraft engine division and obtaining financing with the help of Herbert Quandt, BMW was able to survive. The turnaround was thanks in part to the increasing success of BMW's automotive division. Since the beginnings of its motorcycle manufacturing, BMW periodically introduced single-cylinder models. In 1967, BMW offered the last of these, the R 27. Most of BMW's offerings were still designed to be used with sidecars. By this time sidecars were no longer a consideration of most riders; people were interested in sportier motorcycles. The R 50/2, R 60/2, and R69S marked the end of sidecar-capable BMWs. Of this era, the R 69 S remains the most desirable example of the dubbed "/2" ("slash-two") series because of significantly greater engine power than other models, amongst other features unique to this design.

In 1970 BMW introduced an entirely revamped product line of 500 cc, 600 cc and 750 cc displacement models, the R 50/5, R 60/5 and R 75/5 respectively. The engines were a complete redesign from the older models, producing more power and including electric starting (although the kick-starting feature was still included). The "/5" models were short-lived, however, being replaced by another new product line in 1974. In that year the

BMW R 68

500 cc model was deleted from the lineup and an even bigger 900 cc model was introduced, along with substantial improvements to the electrical system and frame geometry. These models were the R 60/6, R 75/6 and the R 90/6. In 1975 the kick starter was finally eliminated and a supersport model, the BMW R 90 S, was introduced. The R 90 S immediately earned the well deserved title of the best supersport machine available. Today these rare models command high prices in the collector marketplace. Many aficionados of BMW motorcycles view the /5 through /7 lineup as the epitome of classic BMW engineering, though all Airhead models produced through 1995 were roughly similar in terms of owner-friendly maintenance and repair. In addition to "/" or "slash" models, other Airhead models such as the G/S (later, GS) and ST also have dedicated followings within BMW circles while others favor certain earlier models like /5 "toasters." Each has its merits which owners will freely debate with enthusiasm. Later BMW model types such as K-bikes (1983 on) and oilheads (1993 on) included technical innovations that made them more complicated though many owners still elect to service them personally.

In 1977 the product line moved on to the "/7" models. The R 80/7 was added to the line. The R 90 (898 cc) models, "/6" and R 90 S models had their displacement increased to 1,000 cc; replaced by the R 100/7 and the R 100 S, respectively. These were the first liter size (1,000 cc) machines produced by BMW. 1977 was a banner year with the introduction of the first BMW

BMW R 100 RS in the wind tunnel, 1976

production motorcycle featuring a full fairing, the R 100 RS. This sleek model, designed through wind-tunnel testing, produced 70 hp (51 kW) and had a top speed of 200 km/h (124 mph). In 1978, the R 100 RT was introduced into the lineup for the 1979 model year, as the first "full-dress" tourer, designed to compete in this market with the forthcoming Honda Goldwing.

In 1979 the R 60 was replaced with the 650 cc R 65, an entry-level motorcycle with 48 hp (36 kW) that had its very own frame design. Due to its smaller size and better geometrics, front and rear 18-inch (460 mm) wheels and a very light flywheel, was an incredibly well-handling bike that could easily keep up and even run away from its larger brothers when in proper hands on sinuous roads. BMW added a variant in 1982: the R 65 LS, a "sportier" model with a one-fourth fairing, double front disc brakes, stiffer suspension and different carburettors that added 5 hp (4 kW). Not available in the US, was a tuned-down version, the 450 cc R 45 that shared everything with the R 65, but was intended to beat displacement-related licensing taxes in Europe.

In early 1983, BMW introduced a 1000 cc, in-line four-cylinder, water-cooled engine to the European market, the K 100. It was assumed that this new engine would not only be the basis for a new models, but would replace the aging boxer flat twin engine. However, demand for the boxer did not wane with the introduction of this new engine and associated models. And the demand of the new engine models was much less than BMW anticipated, they continued to produce boxer models.

In 1985 BMW produced a 750 cc, three-cylinder version of the new four-cylinder water cooled engine. The 750 cc was counterbalanced, and therefore smoother. The R100RT, boxer powered sport touring bike with a monolever rear suspension was reintroduced in 1987. BMW introduced rear suspension on the K bikes, a double-joined, single-sided swingarm.

In 1986, BMW introduced the world's first electrically adjustable windshield on the K 100 LT. First thought to be an oddity, it has proven to be an important addition to touring motorcycles, is used on numerous BMW models, and has been copied for use on motorcycles by Honda, Moto Guzzi, Kawasaki, and Yamaha, and even on some high-end scooters.

In 1988 BMW introduced ABS on its motorcycles — a first in the motorcycle industry. ABS became standard on all BMW K models. In 1993 ABS was first introduced on BMW's boxer line on the R 1100 RS. It has since spread across nearly all of BMW's shaft-driven motorcycles and even some of its Rotax powered motorcycles.

In 1989, BMW introduced its version of a full-fairing sport bike, the K1. It was based upon the K 100 engine, with four valves per cylinder. Output was near 100 hp (75 kW).

In 1995 BMW ceased production of airhead 2-valve engines and moved its boxer engined line completely over to the newer 4-valve oilhead which were first introduced in 1993.

On 25 September 2004, BMW globally launched a radically redesigned K Series motorcycle, the K 1200 S, containing an all new in-line four-cylinder, liquid-cooled engine featuring 123 kW

(165 hp). The K 1200 S was primarily designed as a Super Sport motorcycle, albeit larger and heavier than the closest Japanese competitors. Shortly after the launch of the K 1200 S, problems were discovered with the new power plant leading to a recall until the beginning of 2005 when corrective changes were put in place. Recently, a K 1200 S set a land speed record for production bikes in its class at the Bonneville Salt Flats, exceeding 174 mph (280 km/h).

In the years after the launch of K 1200 S, BMW has also launched the K 1200 R naked roadster, and the K 1200 GT sports tourer, which started to appear in dealer showrooms in spring (March-June) 2006. All three new K-Series motorcycles are based on the new in-line four-cylinder engine, with slightly varying degrees of power. In 2007 BMW added the K 1200 R Sport, a semi-faired sports touring version of the K 1200 R.

In 2004, bikes with the opposed-twin cylinder "boxer" engine were also revamped. The new boxer displacement is just under 1200 cc, and is affectionately referred to a "hexhead" because of the shape of the cylinder cover. The motor itself is more powerful, and all of the motorcycles that use it are lighter.

The first motorcycle to be launched with this updated engine was the R 1200 GS dual-purpose motorcycle. The R 1200 RT tourer and R 1200 ST sports tourer followed shortly behind. BMW then introduced the 175 kg 105 kW (141 hp) HP2 Enduro, and the 223 kg 100 hp (75 kW) R 1200 GS Adventure, each specifically targeting the off-road and adventure-touring motorcycle segment, respectively. In 2007 the HP2 Enduro was joined by the road-biased HP2 Megamoto fitted with smaller alloy wheels and street tyres.

1990 BMW K1

Notably, the BMW R 1200 C, produced from 1997 to 2004, was BMW Motorcycles first entry into the Cruiser market.

In 2006, BMW launched the R 1200 S, which is rated at 90 kW (121 hp) at 8,250 rpm. April 2007, BMW announced its return to competitive road racing, entering a factory team with a "Sport Boxer" version of the R 1200 S to four 24-hour endurance races. A street version of the R 1200 S Sport Boxer is expected in 2008, rated at 144 hp (107 kW), and weighing 195 kg fully fuelled.

BMW has also paid attention to the F Series in 2006. It lowered the price on the existing F 650 GS and F 650 GS Dakar, and eliminated the F 650 CS to make room in the lineup for the all-new F 800 Series motorcycles. These new F 800 Series motorcycles are powered by a brand new parallel twin engine that is built by Rotax (a Bombardier subsidiary), and a belt drive system that is very similar to the belt drive found on the now defunct F 650 CS. Initially, BMW launched two models of the new F 800 Series, the F 800 S sport bike and the F 800 ST sports tourer; these were followed by an F 800 GS dual-purpose motorcycle.

In October 2006 BMW announced the G series of offroad style motorcycles co-developed with Aprilia, part of the Piaggio group. These are equipped with an uprated single cylinder water cooled 650 cc fuel injected engine producing 53 hp (40 kW), similar to the one fitted to the F 650 GS, and are equipped with chain drive.

BMW K 1200 RS

BMW F 650 GS

BMW F 650 GS

The BMW F 650 GS, introduced in 2000, is a dual-purpose BMW motorcycle, one of the GS on-road/off-road family. It is available in a standard model, and a taller, off-road oriented "Dakar" model, named after the famous Dakar Rally which BMW rider Richard Sainct won on the F650RR in 1999 and 2000. BMW's marketing people refer to this bike as an enduro, but some may feel it is too big and heavy to compete in a sanctioned enduro competition if left in street legal trim; most people would refer to it as a dual-sport or adventure-touring bike.

Its specifications put it in the 650 cc dual-sport class, competing against bikes such as the Kawasaki KLR650, Suzuki DR650, Honda XR650L, KTM LC4 640, Yamaha XT660 and perhaps the Honda Transalp. The standard model is more road-oriented than anything except the Transalp; however, the taller Dakar model can successfully tackle very challenging terrain.

An emergency services specific version of the F 650 GS, fitted with blue lights and sirens, is available from BMW Motorrad's Official and special duty vehicles division.

A specially prepared rally-raid version of the bike was used by Charley Boorman and his team during the 2006 Dakar Rally while filming their documentary Race To Dakar. The single cylinder F 650 GS Dakar model was discontinued in 2007.

In 2008 a completely new F 650 GS model was launched using a 798 cc parallel twin engine. A taller and more powerful F 800 GS was launched using the same engine (with different controller setting).

The F 650 GS has several advanced technology features, with computer-controlled fuel injection, catalytic converter, a Nikasil-lined cylinder, optional ABS and an airbox cleverly designed to exploit the airflow pattern of the bike when in motion. Combined with the bike's high compression ratio and twin spark (from 2004 onwards), fuel economy and reduced emissions exist alongside high power output. The engine is manufactured for BMW by Austrian company Rotax.

Most riders find the F 650 GS more comfortable and less stressed than its competitors at freeway/motorway speeds. The standard model's relatively low seat height make it one of the few 650 cc dual-sports that can be comfortably ridden by riders under 6 ft (182 cm) tall. The F 650 GS has an active aftermarket, with many add-ons and upgrades available. It is slightly more expensive than any of its competitors except the LC4, but that is partially explained by the higher level of standard equipment delivered on the bike. Only the F 650 GS, LC4 and Transalp have hard-luggage available from the manufacturer.

BMW F 650 GS	
PRODUCTION	2000 - Present
MAX SPEED	Ca. 185 km/h
ENGINE TYPE	Two Cylinder
BORE/STROKE	82 mm x 75.6 mm
CAPACITY	798 cc
COMPRESSION RATIO	12.0 : 1
TRANSMISSION	Constant mesh 6-speed
FINAL DRIVE	Endless O-ring chain
VALVE TRAIN	DOHC ; 4 valves per cylinder
CARBURETION	Electronic intake pipe injection
FRONT TIRE	110/80 - 19
REAR TIRE	140/80 - 17
DRY WEIGHT	179 kg
WHEELBASE	2,280 mm
SEAT HEIGHT	820 mm

BMW G 450 X

The BMW G 450 X is a motorcycle exclusively developed for enduro sport and the most demanding competitive races in the world.

The background to the development of such a machine is the knowledge that competition-oriented enduro motorcycles represent a parameter with growth potential in the motorcycle market. Markets such as US, Australia, and Spain, already account for stable sales in the off-road sector and BMW Motorrad wants to help shape and stimulate this and other markets for the long term. Besides, smaller volume off-road motorcycles open the gateway to both popular and professional motor sport as well as offering enough potential to also help the young generation to rediscover the fascination of motorcycles again.

With the launch of the G 450 X, BMW is also pointing the way for the upcoming generation and towards sustainably securing the future of the motorcycle in general. The engine with the latest fuel injection and computer-controlled three-way catalytic converter already meets the vehicle emissions standards which are becoming ever more rigorous today for sports events as well as for road vehicles. With these measures BMW Motorrad is also taking a further step towards public acceptance of enduro sports in the future. The patented, unrivalled technical features of the new BMW G 450 X make it a driving force and contribute to the advancement of technological competition in this segment.

The concept of the BMW G 450 X is based on keeping the mass concentrated as near as possible to the center of gravity and realizing a vehicle configuration that adapts perfectly to the requirements of off-road sport.

The central element of this concept is the amalgamation of the bearing axle of the rear-wheel swing arm with the axis of rotation for the driving pinion, so there is no change in length of the chain on compression and rebound, minimizing the effects of the final drive on the vehicle's response. Further consequences of this unequalled technical approach are a noticeably longer swing arm for the same wheelbase as the competition and, with it, maximum traction bringing concrete advantages in enduro sport with its extreme demands on rider and machine.

BMW G 450 X	
PRODUCTION	2008
MAX SPEED	-
ENGINE TYPE	Single Cylinder 4 stroke
BORE/STROKE	-
CAPACITY	449 cc
COMPRESSION RATIO	-
TRANSMISSION	Constant mesh 5-speed
FINAL DRIVE	Endless O-ring chain
VALVE TRAIN	DOHC ; 4 valves per cylinder
CARBURETION	Electronic intake pipe injection
FRONT TIRE	-
REAR TIRE	-
DRY WEIGHT	120 kg
WHEELBASE	-
SEAT HEIGHT	-

BMW G 650 XCHALLENGE

Powered by a 652cc Single, the new G 650 X represents the lightest-powered machines in the Beemer inventory. A constant on all three versions is the single-cylinder powerplant, now claimed to be 4.5 lbs lighter than the earlier 650 engine. The liquid-cooled mill is claimed to crank out 53 horsepower at 7,000 rpm with torque performance peaking at 44 lb-ft at 5,250 rpm. The 650 thumping away in the heart of the new series was derived from the Single powering the F650GS models, but Motorrad engineers claim modifications to reduce weight and boost performance have upped the power numbers by almost 3 ponies. The juice squeezed out of the Single is delivered to the back wheel via an O-ring chain.

An aluminum radiator remains intact from the F650 design, as well as the gearbox. The stressed-member engine benefits from a lighter crankdrive while a balance shaft remains to ensure smoothness. Other components such as the alternator and starter cover are comprised of magnesium to further reduce weight. Meanwhile, a stainless steel exhaust system, equipped with an oxygen sensor and three-way catalytic converters, ensure the new designs exceed stringent Euro 3 standards.

All three 650s sport a bridge frame constructed out of tubular steel with cast aluminum sections to the side and an aluminum rear frame bolted on. An aluminum-alloy double swingarm rounds out the skeletal design. A 2.1-gallon fuel tank is positioned within the frame triangle beneath the seat, a carryover from the F650 design, and provides an operating range of 155 miles. Strong forged aluminum is used for the foot controls and the sidestands on the Xchallenge and Xmoto.

The big differences between the three variations of the new 650 series come in the spring setting on the 45mm inverted telescopic fork and rear shock, which are calibrated for each machine's particular riding emphasis. Wheel and tire sizes also differ between models, as do the braking configurations.

On the Xchallenge Hard Enduro design, the front fork offers up 10.6 inches of travel and is tuned specific to off-road use, with the unit adjustable for compression and rebound damping. The rear suspension on the Xchallenge is the most different from the other two 650s, as it employs the Air Damping System showcased on the HP2. The rear suspender works like a conventional spring unit but utilizes pressurized air, with the rider able to use a high-pressure pump located underneath the seat to adjust the unit.

The Xchallenge uses a 21-inch front wheel combined with an 18-inch rear, both spoked with aluminum rims and hubs, and accommodate standard 90/90-21 and 140/80-18 off-road meat. A 300mm wave disc teams with a double-piston floating caliper to take care of braking duties up front, while a 240mm rotor (also waved) is grabbed by a single-piston floater out in the back. The Xchallenge is unique to its two 650 siblings with a shorter final drive ratio, via its 15-tooth pinion and 47-tooth sprocket.

BMW G 650 XCHALLENGE

PRODUCTION	2006 - Present
MAX SPEED	Ca. 165 km/h
ENGINE TYPE	Single Cylinder
BORE/STROKE	100 mm x 83 mm
CAPACITY	652 cc
COMPRESSION RATIO	11.5 : 1
TRANSMISSION	Constant mesh 5-speed
FINAL DRIVE	Endless O-ring chain
VALVE TRAIN	DOHC; 4 valve
CARBURETION	Electronic intake pipe injection
FRONT TIRE	90/90 S 21
REAR TIRE	140/80 S 18
DRY WEIGHT	144 kg
WHEELBASE	2,205 mm
SEAT HEIGHT	930 mm

BMW R 1200 GS

For the hearty adventure-touring types, the R1200GS Adventure returns with a few updates aimed at smoothing out the riding experience a bit. The most notable addition is the inclusion of the Enduro ESA (Electronic Suspension Adjustment), that allows the rider to adjust suspension characteristics on the fly.

Other updates include two-piece hand guards, a taller windscreen and a claimed 5% boost in power. An adjustable seat and optional ABS are standard fare on BMW bikes these days but the 8.7-gallon fuel tank, assortment of crash protectors and knobby tires give the Adventure the ability to get the bike deep into unchartered territory and back again.

After four years, the R1200GS gets a few nip/tuck cosmetic enhancements, starting with stainless steel knee covers on both sides of the gas tank, which should provide additional protection from the elements. Both the fork tubes and cylinder head covers have also received some 'sporty' design treatments to go along with the new five-spoke wheels.

In addition to the aesthetic upgrades, the GS also receives the same claimed 5% increase in power as the Adventure. A two-way adjustable handlebar mount allows for a comfy position, whether riding seated or standing up. The Enduro ESA (Electronic Suspension Adjustment) option is also available on the R1200GS.

BMW R 1200 GS

PRODUCTION	2004 - Present
MAX SPEED	Over 200 km/h
ENGINE TYPE	Two Cylinder
BORE/STROKE	100 mm x 83 mm
CAPACITY	1,170 cc
COMPRESSION RATIO	12.0 : 1
TRANSMISSION	Constant mesh 6-speed
FINAL DRIVE	Shaft drive
VALVE TRAIN	SOHC; 4 valve
CARBURETION	Electronic intake pipe injection
FRONT TIRE	110/80 R 19
REAR TIRE	150/70 R 17
DRY WEIGHT	203 kg
WHEELBASE	1,507 mm
SEAT HEIGHT	870 mm

BMW K 1200 R

The K series BMW's have water cooled engines of three (K 75) or four (K 100, K 1100, K 1200) cylinders. Until 2005, the engine was longitudinal, laid out on its left side with the cylinder heads on the left and the crankshaft on the right. It is called the "Flying Brick" because of the appearance of this layout. In 2006, BMW introduced a new 4 cylinder water cooled engine that transverses the chasis and is tilted forward 55 degrees. The BMW K 75, three cylinder, models were produced from 1985 until 1996.

The first K production bike was the K 100, which was introduced in 1983. It was followed by the K 100 RS in 1983, the K 100 RT in 1984, and the K 100 LT in 1986. In 1987, the K 100 (Mark II) was introduced with ABS brakes, the first ever on a motorcycle. In 1988 and until 1993, BMW produced the K1, a full faring version of the K 100 with the new paralever style rear suspension. It had the Bosch Motronic fuel injection system. Initially it cost 20,200 DM. Only 6,900 were produced.

In 1985, the K 75, three cylinder, was introduced. The K 75 C was the first model with this new engine. It was followed by the K 75 S, and the touring version K 75 RT. The last year of production of the K 75 motorcycles was 1996.

In 1991 BMW increased the displacement of the K 100 from 987 cc, and the model designation became the K 1100 (1097 cc). The K 1100 LT was the first with the new engine displacement. In 1992, the K 1100 RS was introduced, ending the 8 year of production of the K 100 models. In 1998 BMW increased the size again to 1170 cc. This upgraded flat four engine appeared in the K 1200 RS. In 2003, a new variation of the K 1200 RS appeared and was designated the K 1200 GT. The K 1200 GT was equipped with hard side cases, larger windshield with electric height adjustment, and a larger fairing. The chasis of the K 1200 RS was extended and strengthened for BMW's luxury touring model the K 1200 LT was still in production in 2007.

The latest K engine is a 1157 cc transverse inline four, announced in 2003 and first seen in the 2005 K 1200 S. The new engine generates a healthy 123 kW (165 hp) but the most striking detail, both visually, and on paper, is its 55 degree forward tilt and 43 cm (17 in) width, giving the bikes a very low center of mass without reducing maximum lean angles. The transverse K 1200 engine is used in K 1200 S, R, R Sport and GT.

BMW K 1200 R

PRODUCTION	2006 - Present
MAX SPEED	Over 200 km/h
ENGINE TYPE	Two Cylinder
BORE/STROKE	79 mm x 59 mm
CAPACITY	1,157 cc
COMPRESSION RATIO	13 : 1
TRANSMISSION	Constant mesh 6-speed
FINAL DRIVE	Shaft drive
VALVE TRAIN	DOHC; 4 valve
CARBURETION	Electronic intake pipe injection
FRONT TIRE	120/70 ZR 17
REAR TIRE	180/55 ZR 17
DRY WEIGHT	215 kg
WHEELBASE	1,580 mm
SEAT HEIGHT	820 mm

BUELL

The current Buell range of bikes

The Buell Motorcycle Company is an American manufacturer based in East Troy, Wisconsin and founded by ex-Harley-Davidson engineer Erik Buell. The company first partnered with Harley-Davidson in 1993, and became a wholly-owned subsidiary of Harley by 1998.

The first Buell motorcycle, the RW750, was built in 1983 purely for competing in the AMA Formula 1 motorcycle road racing championship. At that time, Erik Buell was a top contending privateer motorcycle racer. After completion of the first two RW750 racing machines, one of which was sold to another racing team, the Formula 1 series was cancelled. Buell then turned his focus towards racing-inspired, street-going machines utilizing engines manufactured by Harley. In 1993, Harley-Davidson Incorporated joined in partnership with Buell Motor Company as a 49% stakeholding minority partner and the company formed was renamed "Buell Motorcycle Company". In 1998 Harley purchased majority control of Buell, and it has been a subsidiary ever since. Since then, Buell has utilized modified Harley-Davidson Sportster engines to power their motorcycles.

Most Buell motorcycles use four-stroke air-cooled V-twin engines, originally built from XR1000 Sportster engines. After these were depleted, a basic 1200 Sportster engine was used. In 1995, the engines were upgraded with Buell engineered high performance parts, and further upgraded in 1998.

In 2003, Buell introduced an engine so efficient it passes emissions test requirements through 2008. It does this without the need for catalytic converters, or air injection, as is typical on other modern motorcycles. The new line of Buell XB models also incorporated the industry's first ever Zero Torsional Load (ZTL)

perimeter floating front disc brake system, an "inside-out" wheel/ brake design that puts the brake disc on the outer edge of the wheel, rather than at the hub. This design allows the elimination of significant mass from the front wheel, reducing unsprung weight, and enhances the abilities of the front suspension. Other industry innovations introduced by Buell in the XB lineup were the "fuel in frame technology", and the dual use of the swingarm as an oil tank. Also, all Buell models feature a unique, dual-purpose, muffler mounted below the engine which helps keep mass centralized and maximizes torque through the use of a computer-controlled valve to switch between two exhaust paths as necessary.

Buell designs focus on providing good handling, comfortable riding, easy maintenance, and street-friendly real-world performance. Buell motorcycles are engineered with an emphasis on what they call the "Trilogy of Tech": mass centralization, low unsprung weight, and frame rigidity.

Buell engines are designed to be street-friendly both in fuel efficiency (up to 70 M.P.G. with the Blast), and in power (the 1203cc version produces over 100HP). They are also simple and easy to maintain. Buell two-cylinder engines utilize computer controlled ducted forced air cooling (no radiator or liquid coolant, just a variable speed fan that only activates as required), two valves per cylinder, a single throttle body, zero maintenance hydraulic valve actuation, and zero maintenance gear-driven cams. However, the motor retains a pushrod valvetrain, which was abandoned by Japanese sportsbike manufacturers in the 1970s.

BUELL 1125R

In July, 2007, Buell announced the 1125R, a sportbike which departed from Buell's history of using Harley Davidson style powerplants and tapping into the XBRR racing bike learnings. The Helicon™ engine uses four vertical valves per cylinder, dual over-head cam, liquid-cooled 72 degree V-Twin displacing 1125 cc and producing 146 hp (109 kW). It produces 83 ft·lbf (113 N·m) of peak torque but varies less than 6 ft·lbf (8.1 N·m) of torque from 3,000 to 10,500 rpm. There is a vacuum assist slipper clutch to give predictable drive performance in hard cornering and deceleration and a 6-speed transmission. Although this powerplant is manufactured by BRP-Rotax, Austria, it is in fact designed to detailed Buell specifications.

Said to be designed from the rider down, the Buell trilogy of technology; low unsprung weight, mass centralization and chassis rigidity. The radiator is split and mounted laterally on the bike and air flow managed to lessen the residual heat the rider feels. The split radiator, coupled with the 72 degree twin layout, allows for a more compact and centralized mass making cornering more responsive with the bike's forward weight bias. The 1125R has the greatest lean angle of any Buell. The innovative braking has been further improved by a ZTL2, using four pads and 8 pistons in the caliper with less unsprung weight. The innovative gasoline-in-frame design has been improved further for rigidity while being 10 lb (4.5 kg) lighter than previous designs and an industry first 47 mm inverted forks were matched to improve both squat and dive characteristics.

BUELL 1125R	
PRODUCTION	2007 - Present
MAX SPEED	-
ENGINE TYPE	V-Twin
BORE/STROKE	103 mm x 67.5 mm
CAPACITY	1,125 cc
COMPRESSION RATIO	12.3:1
TRANSMISSION	Six speed straight cut
FINAL DRIVE	Hibrex® Belt drive
VALVE TRAIN	DOHC; 4 valve
CARBURETION	DDFI III fuel injection
FRONT TIRE	120/70 ZR 17
REAR TIRE	180/55 ZR 17
DRY WEIGHT	170 kg
WHEELBASE	1,996 mm
SEAT HEIGHT	775 mm

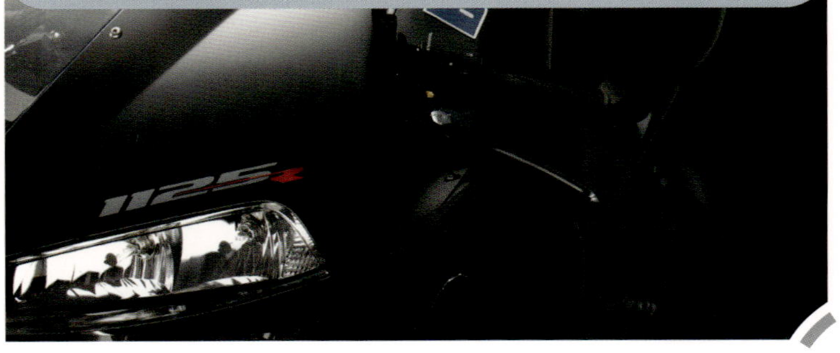

BUELL LIGHTNING CITYX XB9SX

First Introduced in 2005, this urban assault specialist was designed to excel in the metro-madness of crazy cabbies, potholes and back-alley shortcuts.

Geared up for a full frontal assault on the urban jungle, the Lightning® CityX XB9SX is one tough bike. And with its unique see-through flyscreen and airbox cover, you really can see what this bike is made of.

The middleweight Thunderstorm 984 V-Twin revs quicker and higher than the Thunderstorm™ 1203, but still delivers all the torque required to pull away from traffic. For 2008, the Thunderstorm 984 has the new eight-row oil cooler and all of the engine updates applied to the Thunderstorm 1203. Aggressive Pirelli® Scorpion Sync tyres offer good bite on all road surfaces, and a tall Skyline seat with a grippy cover enhances visibility and control. For added style and function, the CityX features wide, Supermoto-style handlebars with deflectors, dual headlight grilles, and frame pucks to fend off minor cosmetic damage.

BUELL LIGHTNING CITYX XB9SX

PRODUCTION	2005 - Present
MAX SPEED	-
ENGINE TYPE	V-Twin
BORE/STROKE	88.9 mm x 79.4 mm
CAPACITY	984 cc
COMPRESSION RATIO	10.0:1
TRANSMISSION	Five-speed, constant mesh
FINAL DRIVE	Hibrex® Belt drive
VALVE TRAIN	OHV; 2 valve
CARBURETION	DDFI III fuel injection
FRONT TIRE	120/70 ZR 17
REAR TIRE	180/55 ZR 17
DRY WEIGHT	177 kg
WHEELBASE	1,950 mm
SEAT HEIGHT	797 mm

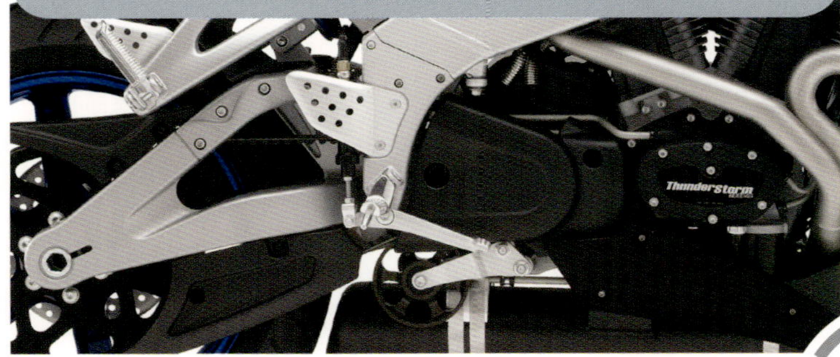

BUELL LIGHTNING LONG XB12SS

In 2005, Buell introduced a Lightning for taller customers and the long haul. The XB12Ss rolls on a 1360mm wheelbase - providing a more relaxed riding position and more room for a passenger. The seat is taller and wider than that of the XB12S, and covers an expanded storage compartment. Larger frame spars give the Lightning Long a fuel-in-frame capacity of 16.7 litres and extended range.

Responsive propulsion is provided by the Buell Thunderstorm™ 1203 V-Twin with Buell InterActive Exhaust. In spite of the longer wheel base the Lightning Long is very much an agile streetfighter with quick and responsive handling.

BUELL LIGHTNING LONG XB12SS

PRODUCTION	2006 - Present
MAX SPEED	-
ENGINE TYPE	V-Twin
BORE/STROKE	88.9 mm x 96.8 mm
CAPACITY	1203 cc
COMPRESSION RATIO	10.0:1
TRANSMISSION	Five-speed, helical gear
FINAL DRIVE	Hibrex® Belt drive
VALVE TRAIN	OHV; 2 valve
CARBURETION	DDFI III fuel injection
FRONT TIRE	120/70 ZR 17
REAR TIRE	180/55 ZR 17
DRY WEIGHT	179 kg
WHEELBASE	2070 mm
SEAT HEIGHT	775 mm

BUELL LIGHTNING XB12S

A direct descendent of the notorious 1996 Buell S1 Lightning, the XB12S is ready to unleash any rider's inner hooligan nature. The tyre-twisting torque of a Thunderstorm 1203 V-Twin is paired with the athletic reflexes of the Buell Intuitive Response Chassis™ (IRC), a combination of throttle response, handling and sexy naked styling that is simply irresistible.

From its flyscreen to its chopped-off tail, the Lightning XB12S sets the standard for streetfighter attitude. Minimal Surlyn bodywork is offered in Midnight Black or Racing Red to offset a new Graphite Grey frame and swingarm, as well as Translucent Amber wheels. The ventilated tail section is Hammertone Silver, while Magnesium Tone is used on the primary cover, cam cover, license plate bracket and front module. The seat has a new textured cover.

BUELL LIGHTNING XB12S

PRODUCTION	2003 - Present
MAX SPEED	-
ENGINE TYPE	V-Twin
BORE/STROKE	88.9 mm x 96.8 mm
CAPACITY	1203 cc
COMPRESSION RATIO	10.0:1
TRANSMISSION	Five-speed, helical gear
FINAL DRIVE	Hibrex® Belt drive
VALVE TRAIN	OHV; 2 valve
CARBURETION	DDFI III fuel injection
FRONT TIRE	120/70 ZR 17
REAR TIRE	180/55 ZR 17
DRY WEIGHT	179 kg
WHEELBASE	1950 mm
SEAT HEIGHT	765 mm

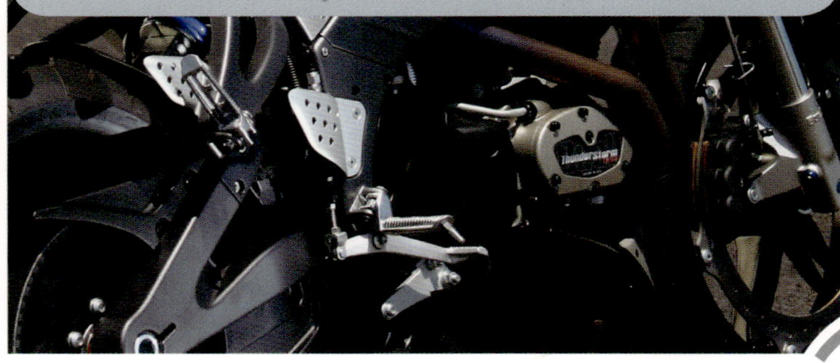

BUELL LIGHTNING XB12SCG

CG stands for 'centre of gravity', and the Lightning XB12Scg keeps it low with a reshaped seat and lowered front and rear suspension, reducing the seat height to 726mm. It's all accomplished with no compromise to handling or performance.

The new Lightning XB12Scgcg offers uncompromised streetfighter performance with a lower seat height. Altered front and rear suspension and a reshaped seat lower the center of gravity and reduce seat height to 28.6 inches - 1.75 inches lower than the standard XB12S. Thrust comes courtesy of a Thunderstorm 1203 V-Twin with Buell InterActive Exhaust and a new, smoother-shifting transmission. Available in Midnight Black or with a new Valencia Orange Translucid airbox cover and flyscreen with Midnight black front fender and chin spoiler, both with high-gloss Translucent Amber wheels.

BUELL LIGHTNING XB12SCG

PRODUCTION	-
MAX SPEED	-
ENGINE TYPE	V-Twin
BORE/STROKE	88.9 mm x 96.8 mm
CAPACITY	1203 cc
COMPRESSION RATIO	10.0:1
TRANSMISSION	Five-speed, constant mesh
FINAL DRIVE	Hibrex® Belt drive
VALVE TRAIN	OHV; 2 valve
CARBURETION	DDFI III fuel injection
FRONT TIRE	120/70 ZR 17
REAR TIRE	180/55 ZR 17
DRY WEIGHT	179 kg
WHEELBASE	1950 mm
SEAT HEIGHT	726 mm

BUELL LIGHTNING XB12STT

The Lightning Super TT brings Supermoto-inspired styling and performance to the street. A narrow-and-flat solo seat makes it easy to shift and set up for the next turn. Wide, flat handlebars provide outstanding leverage, and centred foot

peg placement puts the rider in a commanding position. The clipped, black cast aluminium tail section shows off the 180mm rear tyre. White number plate side panels and flyscreen add a competition look, are designed to be personalised and are easy to change. Fork slider protectors, handlebar deflectors, cosmetic frame protectors, a high front fender and headlight grille protect the Super TT from the abuse of the street. The 1365mm wheelbase gives the rider room to work but retains that trademark Buell agility, while 143mm of suspension travel keeps the Super TT well-planted on rough roads.

BUELL LIGHTNING XB12STT

PRODUCTION	-
MAX SPEED	-
ENGINE TYPE	V-Twin
BORE/STROKE	88.9 mm x 96.8 mm
CAPACITY	1203 cc
COMPRESSION RATIO	10.0:1
TRANSMISSION	Five-speed, helical gear
FINAL DRIVE	Hibrex® Belt drive
VALVE TRAIN	OHV; 2 valve
CARBURETION	DDFI III fuel injection
FRONT TIRE	120/70 ZR 17
REAR TIRE	180/55 ZR 17
DRY WEIGHT	179 kg
WHEELBASE	2080 mm
SEAT HEIGHT	798 mm

BUELL ULYSSES XB12X

Compared with other XB models, Ulysses features long-travel suspension, ample ground clearance and more aggressive tyres that allow it to explore the roughest roads with confidence. But when the surface is smooth and twisty, Ulysses performs like a flickable, agile sportbike. Innovative features and accessories make Ulysses a comfortable, capable long-distance traveler or commuter. This versatile motorcycle can take its rider where ever the road is headed. The 2008 Ulysses will benefit from the newly developed Thunderstorm™ V-Twin engine displacing 1203cc and producing 94 horsepower at 6800 rpm and 77 ft. lbs./104Nm of torque at 6500 rpm. It features an increase in maximum revs, now 7100 rpm up from 6800 rpm, giving this engine a broader powerband and riders the ability to accelerate longer through the gears. The engine also boasts electronic fuel injection and dry-sump oiling.

New 47mm front forks replace the previous 43mm forks. Secured with more robust triple camps, this new front end improves the torsional rigidity of the entire chassis. Fork flex is reduced during hard braking for improved stability and feedback.

Steering sweep lock-to-lock is significantly increased to 74 degrees from 54 degrees to reduce the turning radius in low-speed manoeuvres. New off-set triple clamps maintain the 23.5 degree rake and 122mm trail.

Heated hand grips are now standard equipment on the 2008 Ulysses®. The grips are rated at 18 watts per side on the High setting and 11 watts per side on the Low setting.

New air guides positioned on the tail section of the Ulysses deflect engine heat away from the rider's legs. Modifications to the 1203 Thunderstorm™ V-Twin engine enhance performance, reduce maintenance, and allow a new redline of 7100 rpm.

The instrument panel styling is refreshed and the tachometer reflects the new 7100-rpm redline.

BUELL ULYSSES XB12X	
PRODUCTION	2005 - present
MAX SPEED	-
ENGINE TYPE	V-Twin
BORE/STROKE	88.9 mm x 96.8 mm
CAPACITY	1203 cc
COMPRESSION RATIO	10.0:1
TRANSMISSION	Five-speed, helical gear
FINAL DRIVE	Hibrex® Belt drive
VALVE TRAIN	OHV; 2 valve
CARBURETION	DDFI III fuel injection
FRONT TIRE	120/70 ZR 17
REAR TIRE	180/55 ZR 17
DRY WEIGHT	193 kg
WHEELBASE	2180 mm
SEAT HEIGHT	808 mm

TERRA

MULHACÉN

DERBI

Derbi's origins began with a little bicycle workshop in the village of Mollet near Barcelona, founded in 1922 by Simón Rabasa i Singla The focus remained the repair and hire of bicycles until May 1944 when Singla formed a limited liability company named Bicicletus Rabas with the aim of moving into manufacturing bicycles. The venture proved very successful and in 1946, supported by its profits, work began on a motorised version. More moped than motorcycle, this first model, the 48cc SRS included plunger rear suspension, and a motorcycle type gas tank and exhaust system. The SRS proved so successful it prompted a change in the companies direction, and on November 7, 1950, the company changed its name to the Nacional Motor SA. Just prior to this, at that summers Barcelona Trade Fair, the company unveiled its first real motorcycle, the Derbi 250.

The name Derbi is an acknowledgement of the companies history and is an acronym for DERivados de BIcicletus (derivatives of bicycles).

Unlike Ossa, Bultaco, and Montesa, Derbi successfully met the challenges that followed the Spanish transition to democracy and Spain's entry into the European Community. Simeó Rabasa i Signla passed away in 1988 but the company remained independent until 2001, when it was bought out by the Piaggio group.

1950 Srs 47.75cc

1972 Super Anorcha 49cc

1987 Variant 49cc

1954 4 Cilindros

1956 Super 125

DERBI SENDA DRD EVO 50 SM

The DRD Evo is the latest and most advanced version of the legendary Derbi Senda, the best seller which boasts sales of over half a million motorcycles since it was launched in the early 90s.

Its new and revolutionary configuration, inspired by the most modern, powerful and extreme maxi-motards, allows for a much higher level of road efficiency than seen before. With increased weighting at the front end and a lowered seat (only 835 mm) integrated into the motorcycle, the DRD Evo provides the clearest indication of its innovative position in the 50 motorcycle range.

The Senda DRD Evo incorporates the highest quality components, such as: 17" light alloy wheels, the most powerful and reliable six-speed engine in its category, steel perimeter double-beamed frame, an inverted Marzocchi 40-mm diameter fork and a braking system with radial caliper and Galfer Wave disc brakes on both wheels (320-mm front wheels and 210-mm rear wheels).

Its aggressive and captivating look combined with high impact style, will turn the heads and hearts of all those who love the more extreme and advanced designs.

In 2008, Derbi is destined to stir up the strongest emotions among younger riders, with its strong and defined lines coupled with excellent Supermotard characteristics. DRD Evo: a truly great 50 cc motorcycle that screams high standards and quality.

DERBI SENDA DRD EVO 50 SM	
PRODUCTION	2008
MAX SPEED	-
ENGINE TYPE	Single cylinder 2-stroke
BORE/STROKE	39.86 x 40 mm
CAPACITY	49.9 cc
COMPRESSION RATIO	11.5:1
TRANSMISSION	Six speed
FINAL DRIVE	Chain
VALVE TRAIN	-
CARBURETION	-
FRONT TIRE	100/80 x 17
REAR TIRE	130/70 x 17
DRY WEIGHT	96 kg
WHEELBASE	1,370 mm
SEAT HEIGHT	835 mm

DERBI TERRA ADVENTURE 125

The Terra Adventure 125 was created with the spirit of the numerous adventurers riding their motorcycle in the exploration of new and distant horizons, eager to challenge their very own limits.

The new Terra Adventure is equipped with the 125 4T 4V (Euro3), the most advanced and best performing engine on the market. With this at the heart of the machine, Derbi has underlined its commitment to innovation and reaffirms itself as the absolute pioneer in keeping alive the dream of rally and adventure motorcycles.

The Adventure version offers many significant changes over its predecessor, the Terra 125. It boasts 21" rear spoke wheels and 17" front wheels, Enduro tires, 41-mm larger diameter fork with rod cover, rear mono shock absorber for greater speeds and a bigger aluminum cover for the engine and frame. All these innovations provide this latest Derbi 125 with maximum versatility of use, together with greater comfort and safety, even on the most inaccessible of tracks.

As with the standard model, the defined features and sturdiness of the new Terra Adventure 125 produce a true trail motorcycle, with astonishing results.

The adventurous style of this new model is further illustrated by a new handlebar design (complete with mount for installing a GPS) and handguards connected to an upper fairing, which guarantees excellent aerodynamic protection- therefore more comfort on long rides.

To complete the picture, there is the option of large hard-sided aluminum lateral bags and rear trunk, to make the Derbi Terra Adventure 125 the best 125 cc trail bike and the ideal partner, whatever the adventure.

DERBI TERRA ADVENTURE 125	
PRODUCTION	2008
MAX SPEED	-
ENGINE TYPE	Single cylinder 4-stroke
BORE/STROKE	58 x 47 mm
CAPACITY	124.2 cc
COMPRESSION RATIO	12:1
TRANSMISSION	Six speed
FINAL DRIVE	Chain
VALVE TRAIN	DOHC; 4 valve
CARBURETION	-
FRONT TIRE	90/90 x 21
REAR TIRE	130/90 x 17
DRY WEIGHT	117 kg
WHEELBASE	2,195 mm
SEAT HEIGHT	850 mm

DERBI MULHACÉN CAFÉ 125

The new Derbi Mulhacén Café 125, with stylish and innovative design, makes it the youngest and most sporty motorcycle ever available in the 125 4T market. Fully aware of its renown for style and success in competitions, Derbi created this new model to instill new vitality into this sector.

Equipped with the latest, exclusive dual shaft engine, 4T and 4 valves at 15 HP and a liquid cooling system, the Mulhacén Café 125 is a truly effective combination of sporty style and functionality in an urban motorcycle.

Compared to the standard version, the new Mulhacén 125 boasts 17" aluminum rims, sports tires and a 300 mm rear disc brake with radial caliper. What's more, thanks to the new inverted Marzocchi 40-mm diameter fork , the Mulhacén Café 125 guarantees excellent performance for lovers of sports-style driving even in urban areas. It's ergonomic design makes it a comfortable ride. This combined with its agility in the 125 range, make it the ideal partner in urban areas.

DERBI RAMBLA 125	
PRODUCTION	2008
MAX SPEED	-
ENGINE TYPE	Single cylinder 4-stroke
BORE/STROKE	57x48.6 mm
CAPACITY	124 cc
COMPRESSION RATIO	12.5:1
TRANSMISSION	Automatic
FINAL DRIVE	Chain
VALVE TRAIN	SOHC; 4 valve
CARBURETION	Electronic fuel-injection Euro 3
FRONT TIRE	120/70 x 15
REAR TIRE	130/80 x 15
DRY WEIGHT	144 kg
WHEELBASE	1,360 mm
SEAT HEIGHT	775 mm

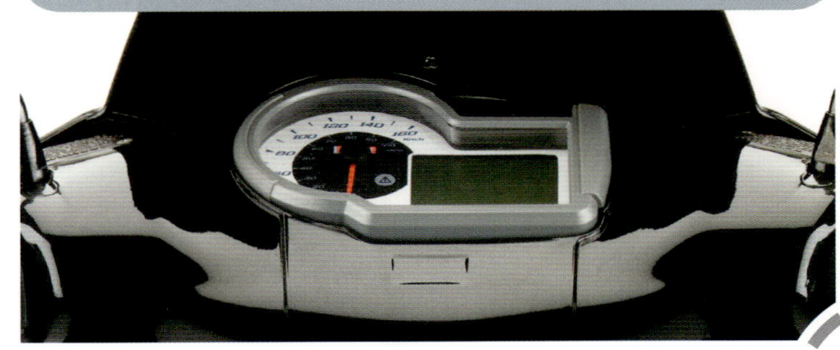

DERBI RAMBLA 250

With technical features chosen with the youngest and most dynamic segment in mind, the Derbi Rambla 250i is the perfect answer to the urban combination of pleasure and practicality.

In the Derbi Rambla, its looks and technology go hand in hand. The joint work of engineers and designers has enabled the creation of a vehicle that seamlessly combines a sophisticated, advanced and comfortable design with high levels of robustness and safety: the definitive sum of Derbi's sporting spirit with all the technical capabilities of the Piaggio engine.

Among its main characteristics is a tough, highly resistant tubular steel chassis guaranteeing maximum stability, large diameter (15") aluminium wheels and wide tyres, which all go to give this vehicle a sporting and very up-to-date image. This performance guarantees comfortable, easy and safe ride in urban traffic, and on all asphalted surfaces. That is why it is the ideal scooter for going to school, travelling to work, for pleasure adventures, and even for going on longer trip.

Other functional aspects worthy of note included a low-slung seat (775mm), suitable for all users, and a generous load capacity. The seat is made of non-slip material which is narrower at the front for improved stability when stationary; its load capacity is explained by a large interior compartment under the seat, where the jet helmet can be stored, a glove compartment, a hook for bags, and a flat platform for an improved posture and an easier ride. Optional accessories for the Derbi Rambla include a windscreen and a storage case that can be fixed on the package carrier with standard side-mounted handles.

The LCD digital/analogue instrument panel gives this vehicle a sporting and up-to-date image. The hot air exhausts behind the front cowl and the 12V connexion (ideal for recharging your cell phone) are other features that facilitate the daily use of this scooter.

The Derbi Rambla 250i features a single cylinder 4-stroke, Euro3 valve engine and 244cc electronic injection and 16.5 kW (22.5 CV) power, with tested efficiency for lower consumption and reduced emissions. It has an extended range, making it a perfect choice for travel outside urban limits. Moreover, the dual front disc (2x260mm) allows for greater braking capacity and safety.

DERBI RAMBLA 250

PRODUCTION	2008
MAX SPEED	-
ENGINE TYPE	Single cylinder 4-stroke
BORE/STROKE	72 x 60 mm
CAPACITY	244.29 cc
COMPRESSION RATIO	11:01
TRANSMISSION	Automatic
FINAL DRIVE	Chain
VALVE TRAIN	SOHC; 4 valve
CARBURETION	Electronic fuel-injection Euro 3
FRONT TIRE	120/70 x 15
REAR TIRE	130/80 x 15
DRY WEIGHT	146 kg
WHEELBASE	1,360 mm
SEAT HEIGHT	775 mm

GILERA

Gilera 175 Super Sport

Gilera's history has its roots in the early part of the twentieth century. The first motorcycle to bear this name, the VT 317, was produced in 1909 by Giuseppe Gilera. In the years following World War I, Gilera produced the 500 cc side valve motorcycles that went on to win major international races. From the mid-Thirties the production of motorcycles with a valve in the crankcase, such as the "Quattro Bulloni 500" and the "Otto Bulloni", began.

The "Rondine", a futuristic cross-racing bike with a four cylinder 500 cc engine, dates to 1936. The motorcycle set several world records (274.181 kph on a flying run in 1937: a record that remained unbeaten for almost two decades) and won Dorino Serafini the 1939 European Championship. After the war Gilera presented the new Saturno 500 and a range of medium-high displacement motorcycles. The four cylinder 500 cc became the new winner: Umberto Masetti was world champion in 1950 and 1952, followed by Geoff Duke (three wins) and Liberati (one title). Gilera also collected six manufacturer's championship wins, three Tourist Trophies, seven Italian titles and an impressive record win by Bruno Francisci at the Milano-Taranto race.

In all Gilera won 44 international Grand Prix titles before its exit from racing in 1957. It was also a strong contender in off-road racing, dominating several International Six Day races. Standard production models included medium-displacement bikes such as the Giubileo, Rossa, Turismo and the Sport. In the higher displacement range, besides the Saturno, the 300 dual cylinder deserves a mention.

In 1969 Gilera became part of the Piaggio Group which undertook a complete reconstruction. It re-launched the historic dual ring brand, transferring it to the production of small and small-to-medium displacements on a range of on and off-road models. Innovations such as the production of the ground breaking 125cc Dual Cylinder Cross motorcycle highlighted the Gilera name again, this time in cross racing.

Gilera 2008 racing team bikes

In the 1980s a new single cylinder four stroke with dual cam distribution was produced, first in standard 350 and 500 cc versions and then in a 600 cc version. This was a winner twice in its class at the Paris-Dakar rally and overall champion at the Pharaohs rally. Gilera's 125cc novelties included the avant-garde powerful "SP O2" and the "CX 125".

In the 1992 and 1993 seasons it returned to Word Championship status in the 250 class.

In 1993 production was transferred to Pontedera and the Gilera brand began to develop sport scooters such as the Runner, an innovative scooter-motorcycle hybrid. In all, nearly 700,000 Gilera branded vehicles have been produced since 1993.

Multi-gear small motorcycles include the enduros H@K, Surfer and the "supermotard" GSM. In 1998 Gilera returned to its motorcycling origins with the 125cc four stroke cruiser Coguar. The year 2000 marked the launch of the revolutionary DNA, a "naked" motorcycle with an automatic engine that pushes the scooter and motorcycle concepts even closer together.

In 2001 the double-ring brand returned to the Grand Prix. With the young rider Manuel Poggiali (a native of the Republic of San Marino), Gilera was one of the top players in the 125cc class, with Poggiali heading the classifications. On the 20th May 2001 on the Le Mans circuit Poggiali registered a breathtaking victory, arriving first at the finish line. It had been 38 years since the last Gilera victory, when John Hurtle triumphed in the 500cc class at the Grand Prix in Assen, Holland. At the end of an exciting season, having notched up another two victories at the Portugal and Valencia Grand Prix and a total of 11 podium placements, Manuel Poggiali was 2001 World Champion in the 125cc class.

It was a victory that had been awaited for 44 years. Gilera has been truly back in the race.

GILERA FUOCO 500IE

The Gilera Fuoco's innovative looks get its rider noticed. The front end of the Gilera Fuoco 500ie is attention grabbing: its decisive shape expresses strength and character and hints at the incredible performance of the all-new 492cc single cylinder engine. The two-wheel frontal is marked by a steel tube bumper with metal mesh inserts that give the vehicle a rugged look. The front design protects the rider and the mechanicals while putting the unique front suspension technology on display. Other wicked touches are the sleek 'naked' metal handlebar and black ten-spoke wheel rims. The five-lamp headlight unit is not only striking to look at but highly effective as well; the two biggest lamps are equipped with off-road-style shockproof covers. The front fairing and front shield fairing offer remarkable aerodynamic protection. There is a wide, comfortable footrest panel behind the front shield fairing.

The tail design is stark and minimalist, a mix of plastics and tubes that picks up on the styling of the frontal with a very useful rear rack, while the wide, comfortable seat ensures total comfort for both rider and passenger thanks to its ergonomic design and negligible height difference between the two parts of the seat.

The steel tube frame conceals the powerful, reliable new Master 500ie double ignition engine, a 4 valve, 4 stroke unit with electronic injection and liquid cooling. The capacity of the new Master engine has been upped to 492 cc to obtain maximum power of 40 hp at 7,250 rpm and maximum torque of over 42 Nm at 5,500 rpm. The introduction of the twin spark system has also made it possible to optimise combustion inside the cylinder, with a reduction in noise and gas emissions. The result is a smooth, high-performance engine, very torquey at low and medium range rpm, that takes the Gilera Fuoco 500ie to a top speed of nearly 145 km/h while fully respecting Euro 3 norms thanks to the advanced closed loop injection circuit with a Lambda sensor and three-way catalytic converter in the exhaust pipe.

The engine's exuberance is skilfully managed by the sophisticated running gear. The innovative parallelogram front suspension's tilt mechanism is composed of four cast aluminium arms, with four hinges fixed to the central tube and two guide tubes on either side of the parallelogram, connected to the arms via suspension pins and ball bearings. This means that the Gilera Fuoco 500ie is as easy to ride as a traditional scooter, while its incredible stability, especially when cornering and braking, comes from its two front wheels.

Standard equipment includes an electro-hydraulic front suspension locking system that keeps the Fuoco 500ie upright without a central stand. This makes it extremely easy to park anywhere. What's more, there's no need to put your feet on the ground to keep your balance when stopped at a traffic light.

Optimal rear end stability is guaranteed by a 14" rear wheel with a generous 140/70 tyre, while three 240mm disk brakes with dual-piston calipers ensure fast, efficient braking.

GILERA FUOCO 500IE

PRODUCTION	2007 - present
MAX SPEED	143 Km/h
ENGINE TYPE	Single cylinder 4-stroke
BORE/STROKE	94mm x 71mm
CAPACITY	492.7 cc
COMPRESSION RATIO	11:1
TRANSMISSION	Automatic
FINAL DRIVE	-
VALVE TRAIN	SOHC; 4 valve
CARBURETION	Electronic fuel-injection Euro 3
FRONT TIRE	2 x 120/70 x 12
REAR TIRE	130/80 x 14
DRY WEIGHT	244 kg
WHEELBASE	1,550 mm
SEAT HEIGHT	785 mm

GILERA GP800

Everything about the Gilera GP 800 has been designed, developed and made to rule the road. Each individual stylistic choice and technical solution speaks of a revolutionary machine, conceived around the first ever 90° V twin scooter engine, and designed to dominate the world of scooters. The new 839.3 cc engine features liquid cooling, Magneti Marelli electronic fuel injection, a single cam, 4 valves per cylinder and twin spark ignition. It develops a mighty 75 bhp at the crank with a torque of 76.4 Nm, 95% of which is already available at 3,500 rpm. These specifications are totally unique in the scooter world. On the road they translate into an extraordinary reserve of constantly available power and a uniquely smooth ride. Thanks to CVT (continuously variable transmission) with engine braking effect, urban commuting is effortless and relaxing, with the ideal ratio automatically selected. But when the road leaves the town and opens up, the GP 800's V twin engine really comes into its own, placing Gilera's flagship scooter in a class of its own.

The Gilera GP 800 engine features dry sump lubrication, and is secured rigidly to the frame by vibration-damping rubber mountings. Final drive can therefore be entrusted to a robust but sophisticated, Regina Z-ring sealed chain.

The exhaust system is totally decoupled from the frame to minimise vibrations. This 2-in-1 system terminates in an aggressively styled silencer with a black heat shield that adds plenty of character to the styling of the GP 800, and emphasises its elegant but sporting nature.

The amazing performance of the GP 800's V twin engine have been achieved in full respect of the environment. Emissions have been minimised by an advanced closed loop injection system with Lambda oxygen sensor and an exhaust with three-way catalytic converter. Fully integrated control of the electronic ignition system also ensures that the GP 800 conforms to strict Euro 3 standards.

The new twin cylinder power plant has been designed, developed and manufactured entirely in Pontedera, the historical home of Piaggio and one of the world's most advanced engine development and production centres. This engine is born from the challenge that the Piaggio Group has taken on: to offer an ever more sophisticated customer a product that guarantees performance, reliability and safety while also leading the field in respect for the environment.

GILERA GP800

PRODUCTION	2007 - present
MAX SPEED	200 Km/h
ENGINE TYPE	V-twin 4-stroke
BORE/STROKE	88mm x 94mm
CAPACITY	839.3 cc
COMPRESSION RATIO	10.5:1
TRANSMISSION	CVT
FINAL DRIVE	Chain
VALVE TRAIN	SOHC; 4 valve
CARBURETION	Electronic fuel-injection
FRONT TIRE	120/70 R 16
REAR TIRE	160/60 R 15
DRY WEIGHT	245 kg
WHEELBASE	1,593 mm
SEAT HEIGHT	780 mm

GILERA NEXUS 125

The Nexus 125's aggressive looks suggest sport equipment and in fact its technical features are worthy of a higher-displacement scooter. The Nexus 125 uses exactly the same frame as the 250cc, with a strong, reliable double shock absorber at the rear to enhance lightness and build simplicity and a traditional fork with 35mm shafts in front.

The braking system is made up of a 260 mm disc with a floating two-piston caliper in front and a 240 mm disc gripped by a caliper with opposite pistons on the rear wheel.

The 14" wheel rims mount a 120/70 tyre in front and a sizeable 140/60 at the rear, a choice that suits the sporty nature of the smallest Nexus.

The instrumentation and trip controls are the same ones used on the 250 and 500cc, further proof of an uncompromising design philosophy.

GILERA NEXUS 125	
PRODUCTION	2007 - present
MAX SPEED	103 Km/h
ENGINE TYPE	Piaggio LEADER
BORE/STROKE	57mm x 46mm
CAPACITY	124 cc
COMPRESSION RATIO	12:1
TRANSMISSION	CVT
FINAL DRIVE	-
VALVE TRAIN	SOHC; 4 valve
CARBURETION	Electronic fuel-injection
FRONT TIRE	120/70-14
REAR TIRE	140/60-14
DRY WEIGHT	186 kg
WHEELBASE	1,530 mm
SEAT HEIGHT	810 mm

GILERA STORM

The new Gilera Storm is packed with personality. The front is characterised by streamlined lines and a leg shield that extends forwards over the wheel to act as front mudguard. Air intakes just under the headlight cluster enhance the scooter's sporty appearance and optimise air flow aerodynamics, making the Gilera Storm even more aggressive. The handlebars are mounted well above the leg shield, and extend from a tinted nose fairing that gives the Storm an aggressive look and houses flush-fitting, clear plastic direction indicators.

The sporty, energy-packed soul of the Storm is enhanced by racing graphics and characterised by attractive details like the mirrors, nose fairing, platform, and exhaust. The shiny black wheels even feature a red trim line in keeping with Gilera's legendary racing tradition. The sleek rear of the scooter, with its characteristic lateral air intakes, embodies typical Gilera racing spirit.

Designed for everyday life on the toughest of all circuits – our towns and cities – the Storm provides a comfortable ride even two up. Added passenger safety is provided by a grab strap on the luxurious long seat while passenger comfort is ensured by practical folding footrests.

Even the design of the instruments pays homage to the sporting soul of Gilera. The round instruments (speedometer, rev counter and fuel gauge) feature red digits on a black background and are elegantly trimmed in a combination of matt black and satin finished aluminium. The instrumentation comes complete with fuel reserve and low oil warning lights, turn, dipped and main beam indicators.

The new Gilera Storm stands out for the reliability and safety afforded by its advanced chassis. The frame is made from high strength steel tube. The super-sporty, 12" five spoke wheels are both fitted with 120/70 tubeless tyres with road-going tread patterns. Powerful, safe braking is provided by a 190 mm front disc and a reliable 110 mm drum brake at the rear. There can be no doubt about the strength and reliability of the suspension either: the upside down hydraulic fork boasts 30 mm stanchions and the hydraulic monoshock at the rear is extremely robust in design.

The Gilera Storm is powered by a reliable, modern, 50 cc air cooled two stroke engine that conforms to Euro2 standards, the strictest emission control standards required of 50cc two wheelers. The Hi-Per2 engine delivers plenty of performance for riders seeking a powerful but young and easy-to-use engine. Excellent engine flexibility makes the new Storm the perfect companion around town, where any journey is a continuous succession of "stop and go" manoeuvres.

GILERA STORM	
PRODUCTION	2007 - present
MAX SPEED	-
ENGINE TYPE	Single cylinder 2 stroke
BORE/STROKE	40mm x 39.3mm
CAPACITY	49 cc
COMPRESSION RATIO	-
TRANSMISSION	CVT
FINAL DRIVE	-
VALVE TRAIN	-
CARBURETION	Electronic fuel-injection
FRONT TIRE	120/70-12
REAR TIRE	120/60-12
DRY WEIGHT	81 kg
WHEELBASE	1,280 mm
SEAT HEIGHT	810 mm

HARLEY DAVIDSON

Harley-Davidson Motor Company is an American manufacturer of motorcycles based in Milwaukee, Wisconsin. The company sells heavyweight (over 750 cc) motorcycles designed for cruising on the highway. Harley-Davidson motorcycles (popularly known as "Harleys") have a distinctive design and exhaust note. They are especially noted for the tradition of heavy customization that gave rise to the chopper-style of motorcycle.

Harley-Davidson attracts a loyal brand community, with licensing of the Harley-Davidson logo accounting for almost 5% of the company's net revenue ($41 million in 2004). In 2003, the Buell Motorcycle Company became a wholly-owned subsidiary of Harley-Davidson, the same year that the Motor Company celebrated its 100th birthday. The Motor Company supplies many American police forces with their motorcycle fleets.

The company considers 1903 to be its year of founding, though the Harley-Davidson enterprise could be considered to have started in 1901 when William S. Harley, age 21, drew up plans for a small engine that displaced 7.07 cubic inches (116 cc) and had four-inch (102 mm) flywheels. The engine was designed for use in a regular pedal-bicycle frame.

Over the next two years Harley and his boyhood friend Arthur Davidson labored on their motor-bicycle using the northside machine shop at the home of their friend, Henry Melk. It was finished in 1903 with the help of Arthur's brother, Walter Davidson. Upon completion the boys found their power-cycle unable to conquer Milwaukee's modest hills without pedal assistance. Will Harley and the Davidsons quickly wrote off their first motor-bicycle as a valuable learning experiment.

Work was immediately begun on a new and improved second-generation machine. This first "real" Harley-Davidson motorcycle had a bigger engine of 24.74 cubic inches (405 cc) with 9-3/4 inch flywheels weighing 28 pounds. The machine's advanced loop-frame pattern was similar to the 1903 Milwaukee Merkel motorcycle (designed by Joseph Merkel, later of Flying Merkel fame.) The bigger engine and loop-frame design took it out of the motorized-bicycle category and would help define what a modern motorcycle should contain in the years to come. The boys also received help with their bigger engine from outboard motor pioneer Ole Evinrude, who was then building gas engines of his own design for automotive use on Milwaukee's Lake Street.

The prototype of the new loop-frame Harley-Davidson was assembled in a 10- by 15-foot (3 by 5 meter) shed in the Davidson family backyard. Most of the major parts, however, were made elsewhere, including some probably fabricated at the West Milwaukee railshops where oldest brother William A. Davidson was then toolroom foreman. This prototype machine was functional by 8 September 1904 when it competed in a Milwaukee motorcycle race held at State Fair Park. It was ridden by Edward Hildebrand and placed fourth. This is the first documented appearance of a Harley-Davidson motorcycle in the historical record.

In January 1905, small advertisements were placed in the "Automobile and Cycle Trade Journal" that offered bare Harley-Davidson engines to the do-it-yourself trade. By April, complete motorcycles were in production on a very limited basis. That year the first Harley-Davidson dealer, Carl H. Lang of Chicago, sold

three bikes from the dozen or so built in the Davidson backyard shed. (Some years later the original shed was taken to the Juneau Avenue factory where it would stand for many decades as a tribute to the Motor Company's humble origins. Unfortunately, the first shed was accidentally destroyed by contractors in the early 1970s during a clean-up of the factory yard.)

In 1906, Harley and the Davidsons built their first factory on Chestnut Street (later Juneau Avenue). This location remains the Motor Company's corporate headquarters today. The first Juneau Avenue plant was a modest 40 by 60-foot (18 m) single-story wooden structure. That year around 50 motorcycles were produced.

In 1907, William S. Harley graduated from the University of Wisconsin-Madison with a degree in mechanical engineering. That year additional factory expansion came with a second floor and later with facings and additions of Milwaukee pale yellow ("cream") brick. With the new facilities production increased to 150 motorcycles in 1907. That September a milestone was reached when the fledgling company was officially incorporated. They also began selling their motorcycles to police departments around this time, a tradition that continues today.

Production in 1905 and 1906 were all single-cylinder models with 26.84 cubic inch (440 cc) engines but as early as February of 1907 a prototype model with a 45-degree V-Twin engine was displayed at the Chicago Automobile Show. Although shown and advertised, very few dual cylinder V-Twin models were built between 1907 and 1910. These first V-Twins displaced 53.68 cubic inches (880 cc) and produced about 7 horsepower (5 kW). This gave about double the hill-climbing power of the first singles. Top speed was about 60 mph (97 km/h). Production jumped from 450 motorcycles in 1908 to 1,149 machines in 1909.

The success of Harley-Davidson (along with Indian's success) had attracted many imitators. By 1911 some 150 makes of motorcycles had already been built in the United States - although just a handful would survive the 1910s.

In 1910, the famed "Bar & Shield" logo was used for the first time. It is trademarked at the U.S. Patent office one year later.

In 1911, an improved V-Twin model with mechanically operated intake valves was introduced. (Earlier V-Twins had used "automatic" intake valves that opened by engine vacuum). Displacing 49.48 cubic inches (810 cc), the 1911 V-Twin was actually smaller than earlier twins, but gave better performance. After 1913 the majority of bikes produced by Harley-Davidson would be V-Twin models.

By 1913, the yellow brick factory had been demolished and on the site a new 5-story structure of reinforced concrete and red brick had been built. Begun in 1910, the red brick factory with its many additions would take up two blocks along Juneau Avenue and around the corner on 38th Street. Despite the competition, Harley-Davidson was already pulling ahead of Indian and would dominate motorcycle racing after 1914. Production that year swelled to 16,284 machines.

When America is plunged into World War II. Production of civilian motorcycles is almost entirely suspended in favor of military production. The Service School is converted back to the Quartermasters School for the training of military mechanics.

Harley-Davidson receives the first of its four Army-Navy "E" Awards for excellence in wartime production. Overseas, many American servicemen get their first exposure to Harley-Davidson motorcycles, something they would not forget when they would return stateside.

World War II ends, and Harley-Davidson has produced almost 90,000 WLA models for military use. Wasting no time, production of civilian motorcycles resumes in November.

As part of war reparations, Harley-Davidson acquired the design of a small German motorcycle, the DKW RT125 which they adapted, manufactured, and sold from 1947 to 1966. Various models were made, including the Hummer from 1955 to 1959, but they are all colloquially referred to as "Hummers" at present. BSA in the United Kingdom took the same design as the foundation of their BSA Bantam.

In 1960, Harley-Davidson consolidated the Model 165 and Hummer lines into the Super-10, introduced the Topper scooter, and bought fifty percent of Aeronautica Macchi's motorcycle division. Importation of Aermacchi's 250 cc horizontal single began the following year. The bike bore Harley-Davidson badges and was marketed as the Harley-Davidson Sprint.

After the Pacer and Scat models were discontinued at the end of 1965, the Bobcat became the last of Harley-Davidson's American-made two-stroke motorcycles. The Bobcat was manufactured only in the 1966 model year.

Harley-Davidson's entry in the lightweight two-stroke market for 1967 was the M-65, built by Aermacchi and offered in base form with a semi-step thru frame and tank and as the M-65S (Sport) with a larger tank (later used on the 1968 Rapido).

The company re-entered the 125 cc two-stroke market in 1968 with the introduction of the Aermacchi-built Rapido, a 125 cc bike to replace the American-made 2-stroke bikes.

The engine of the Sprint was increased to 350 cc in 1969 and would remain that size until 1974, when it was replaced by the 250 cc two-stroke SX.

Harley-Davidson purchased full control of Aermacchi's motorcycle production in 1974 and continued making two-stroke motorcycles there until 1978, when they sold the facility to Cagiva.

In 1969, American Machinery and Foundry (AMF) bought the company, streamlined production, and slashed the workforce.

In 1981, AMF sold the company to a group of thirteen investors led by Vaughn Beals and Willie G. Davidson for $80 million. Inventory was strictly controlled using the Just In Time system.

The "Sturgis" model, boasting a dual belt-drive, was introduced. By 1990, with the introduction of the "Fat Boy", Harley once again became the sales leader in the heavyweight (over 750 cc) market. At the time of the Fat Boy model introduction a story rapidly spread that its silver paint job and other features were inspired by the World War II American B-29 bomber; and that the Fat Boy name was a combination of the names of the atom bombs (Fat Man and Little Boy) that were dropped on Nagasaki and Hiroshima respectively. However, the Urban Legend Reference Pages lists this story as an urban legend.

The VRSCA V-Rod® is introduced for the 2002 model year. Inspired by the VR-1000 racing motorcycle, the V-Rod is Harley-Davidson's first motorcycle to combine fuel injection, overhead cams and liquid cooling, and delivers 115 horsepower.

In 2003, more than 250,000 people come to Milwaukee for the final stop of the Open Road Tour and the Harley-Davidson 100th Anniversary Celebration and Party.

HARLEY DAVIDSON SPORTSTER

Introduced in 1957, the Sportster is the longest-running model family in the Harley-Davidson lineup. They were conceived as racing motorcycles, and were popular on dirt and flat-track race courses through the 1960s and '70s. Smaller and lighter than the other Harley models, contemporary Sportsters make use of 883 or 1,200 cc Evolution engines and, though often modified, remain similar in appearance to their racing ancestors.

Up until the 2003 model year, the engine on the Sportster was rigidly mounted to the frame. The 2004 Sportster had a new frame accommodating a rubber-mounted engine. Although this made the bike heavier and reduced the available lean angle, it reduced the amount of vibration transmitted to the frame and the rider. The rubber mounted engine provides a significantly smoother ride for rider and passenger. For a bike which isn't really thought of for long rides or trips, the smoother ride allows for this opportunity.

In the 2007 model year, Harley Davidson celebrated the 50th anniversary of the Sportster and produced a collectors' edition called the XL50 1200 Custom, of which only 2000 were made for sale world wide. Each motorcycle was individually numbered and came in one of two colours, Mirage Pearl Orange or Vivid Black. Also in 2007, electronic fuel injection was introduced to the Sportster family. 2007 was also the first year Harley Davidson made the Nightster.

HARLEY DAVIDSON SPORTSTER NIGHTSTER

PRODUCTION	2008 - present
MAX SPEED	-
ENGINE TYPE	V-twin 4 stroke
BORE/STROKE	88.9mm x 96.8mm
CAPACITY	1200 cc
COMPRESSION RATIO	9.7:1
TRANSMISSION	5-speed
FINAL DRIVE	Belt
VALVE TRAIN	SOHC; 4 Valve
CARBURETION	Electronic fuel-injection
FRONT TIRE	100/90-19
REAR TIRE	150/80-16
DRY WEIGHT	251 kg
WHEELBASE	1,510 mm
SEAT HEIGHT	676 mm

HARLEY DAVIDSON DYNA

Design work began on the replacement for the FXR chassis shortly after the first FXR bikes were offered. The Dyna chassis was introduced in 1991 with a limited-production FXDB Sturgis model. The engine mounting system was more vibration-resistant than that of the FXR.

Between the 1991 introduction of the Dyna chassis and the end of the 1994 model year, all Dyna models had a 32° rake. In 1995 the FXD Dyna Super Glide and the FXDS-Conv Dyna Glide Convertible were introduced. These Dynas had a 28° rake and replaced the FXR Super Glide and the FXRS-Conv Low Rider Convertible, which were the last FXR models in regular production.

In 2006, a new Dyna chassis was introduced. In the same year, the base FXDI Super Glide became a single-seat motorcycle, the FXDBI Street Bob, a minimal, single seat Dyna Glide motorcycle was added to the lineup, the limited edition FXDI35 35th Anniversary Super Glide was offered, and the FXDX Super Glide Sport was discontinued.

In 2007, the Twin Cam 88 engine was replaced by the 1584cc Twin Cam 96 engine across the Harley-Davidson Big Twin lineup, including the FXD series. The FXDC Super Glide Custom was introduced in the same year.

HARLEY DAVIDSON DYNA STREET BOB

PRODUCTION	2006 - present
MAX SPEED	-
ENGINE TYPE	Twin Cam 4 stroke
BORE/STROKE	88.9mm x 96.8mm
CAPACITY	1584 cc
COMPRESSION RATIO	9.2:1
TRANSMISSION	6-speed
FINAL DRIVE	Belt
VALVE TRAIN	DOHC; 4 Valve
CARBURETION	Electronic fuel-injection
FRONT TIRE	100/90-19
REAR TIRE	160/70-17
DRY WEIGHT	290 kg
WHEELBASE	1,630 mm
SEAT HEIGHT	655 mm

HARLEY DAVIDSON SOFTAIL

The term softail refers to motorcycles and bicycles that feature a moveable rear suspension system with springs or shock absorbers to absorb bumps. On motorcycles, the shock absorbers or springs are hidden underneath out of view to give the appearance of a hard-tail or rigid frame. The word softail is a registered trademark of Harley-Davidson motorcycles coined with the release of the FXST Softail in 1984. Since then, the word has expanded to include other motorcycles with hidden rear suspensions as well as bicycles incorporating a rear suspension.

In Harley-Davidson motorcycles, the softail frame is designed to look like the hardtail bikes of the past, while still offering the comfort of rear suspension. The shock absorbers are positioned along the axis of the motorcycle, tucked away under the transmission.

There are several Harley-Davidson models with the Softail frame, including the Softail Standard, Custom, Springer Softail, Heritage Softail, Heritage Springer, Night Train, Deluxe, Deuce, and Fatboy. These motorcycles have the same engine, transmission and frame with the exception of the Deuce, which has a 2" backbone stretch, but differ in the choice of fork, wheels and accessories.

The Softail range is the only range of Harley-Davidson motorcycle that offers a choice of front suspensions within the range. The Sportster and Dyna models use the thin X-type telescopic forks, the VRSC models use a fork unique to the VRSC line, and the Touring models use the fat FL-type telescopic forks. Softails models may use either the X-type or the FL-type, or may use the Springer leading link forks that have only been used on Softail models so far. The FXST designation is used when the X-type fork is used or when the Springer fork is used with a 21" wheel, while the FLST designation is used when the FL-type fork is used or when the Springer fork is used with a 16" wheel.

HARLEY DAVIDSON SOFTAIL FAT BOY

PRODUCTION	2006 - present
MAX SPEED	-
ENGINE TYPE	Twin Cam 4 stroke
BORE/STROKE	95.3mm x 111.1mm
CAPACITY	1584 cc
COMPRESSION RATIO	9.2:1
TRANSMISSION	6-speed
FINAL DRIVE	Belt
VALVE TRAIN	DOHC; 4 Valve
CARBURETION	Electronic fuel-injection
FRONT TIRE	140/75-17
REAR TIRE	200/55-17
DRY WEIGHT	313 kg
WHEELBASE	1,635 mm
SEAT HEIGHT	645 mm

HARLEY DAVIDSON VSRC

Introduced in 2001, the VRSC family bears little resemblance to Harley's more traditional lineup. Competing against Japanese and American muscle bikes and seeking to expand its market appeal, the "V-Rod" makes use of an engine developed jointly with Porsche that, for the first time in Harley history, incorporates fuel injection, overhead cams, and liquid cooling. The V-Rod is visually distinctive, easily identified by the 60-degree V-Twin engine, the radiator and the hydroformed frame members that support the round-topped air cleaner cover. Based on the VR-1000 racing motorcycle, it continues to be a platform around which Harley-Davidson builds drag-racing competition machines. The V-Rod has gathered an enthusiastic following in the U.S., Europe and Australia, and an annual Rally at the Kansas City production facility has been organized by Max Millender and the members of a 21,000+ member strong internet discussion forum www.1130cc.com. Bill Davidson has presented Mr Millender with a signed airbox cover to recognize the contribution the forum has made to the VRSC platform which continues to evolve with models like the Night Rod Special, or VRSCDX.

In 2008, Harley added anti-lock braking systems as a factory installed option on all VRSC models. Harley also increased the displacement of the engine from 1130 cc to 1250 cc and added a slipper clutch as standard equipment.

HARLEY DAVIDSON VSRC NIGHT ROD

PRODUCTION	2001 - present
MAX SPEED	-
ENGINE TYPE	V-twin 4 stroke
BORE/STROKE	105mm x 72mm
CAPACITY	1250 cc
COMPRESSION RATIO	11.3:1
TRANSMISSION	5-speed
FINAL DRIVE	Belt
VALVE TRAIN	DOHC; 4 Valve
CARBURETION	Electronic fuel-injection
FRONT TIRE	120/70-19
REAR TIRE	240/40-18
DRY WEIGHT	292 kg
WHEELBASE	1,715 mm
SEAT HEIGHT	640 mm

HARLEY DAVIDSON VSRC

HARLEY DAVIDSON TOURING

The touring family, also known as "dressers", includes three Road King models, and five Glide models offered in various trim. The Road Kings have a "retro cruiser" appearance and most models are equipped with a large clear windshield. Road Kings are reminiscent of big-twin models from the 1940s and '50s. Glides can be identified by their full front fairings. Most Glides sport a unique fairing referred to as the "Batwing" due to its unmistakable shape. The Road Glide has a different front end, referred to as the "Sharknose". The Sharknose includes a unique, dual front headlight. Touring models are distinguishable by their large luggage, rear coil-over air suspension and are the only models to offer full fairings with Radios/CBs. All touring models use the same frame, first introduced with a Shovelhead motor in 1980, and carried forward with only modest upgrades to this day. The frame is distinguished by the location of the steering head in front of the forks and was the first H-D frame to rubber mount the drivetrain to isolate the rider from the vibration of the big V-twin. Although all touring models weigh in excess of 800 lb (360 kg)., they are remarkably easy to handle at low speeds and high, and give the most comfortable and relaxing ride of any Harley. The frame was modified for the 1994 model year when the oil tank went under the transmission and the battery was moved inboard from under the right saddlebag to under the seat. In 1997, the frame was again modified to allow for a larger battery under the seat and to lower seat height. In 2007, Harley introduced a the 96 cubic inch motor, as well the 6 speed transmission to give the rider better speeds on the highway.

HARLEY DAVIDSON TOURING ELECTRA GLIDE

PRODUCTION	1965 - present
MAX SPEED	-
ENGINE TYPE	Twin Cam 4 stroke
BORE/STROKE	95.3mm x 111.1mm
CAPACITY	1584 cc
COMPRESSION RATIO	9.2:1
TRANSMISSION	6-speed
FINAL DRIVE	Belt
VALVE TRAIN	DOHC; 4 Valve
CARBURETION	Electronic fuel-injection
FRONT TIRE	D402F MT90B16
REAR TIRE	D402 MU85B16
DRY WEIGHT	332 kg
WHEELBASE	1,610 mm
SEAT HEIGHT	693 mm

CBR
HONDA
TOKICO
ROCKET

HONDA

Mr. Honda as a young man.

Mr. Honda on the assembly line.

Company founder Soichiro Honda, after working at Art Shokai, developed his own design for piston rings in 1938. He attempted to sell them to Toyota and after two years of work he won a contract with them. He constructed a new facility to supply Toyota, but soon after, during World War II, the Honda piston manufacturing facilities were almost completely destroyed.

Soichiro Honda created a new company with what he had left. The Japanese market was crippled by World War II; his country was starved of money and fuel, but was still in need of basic transportation. Honda, utilizing his manufacturing facilities, attached an engine to a bicycle which created a cheap and efficient method of transport. He gave his company the name Honda Giken Kgy Kabushiki Kaisha which translates to Honda Research Institute Company Ltd. Despite its grandiose name, the first facility bearing that name was a simple wooden shack where Mr. Honda and his associates would fit the engines to bicycles. The official Japanese name for Honda Motor Company Ltd. remains the same in honor of Soichiro Honda's efforts. On 24 September 1948 the Honda Motor Co. was officially founded in Japan.

Honda's first motorcycle to be put on sale was the 1947 A-Type (one year before the company was officially founded). However, Honda's first full-fledged motorcycle on the market was the 1949 Dream D-Type. It was equipped with a 98cc engine producing around 3 horsepower (2.2 kW). This was followed by other highly popular scooters throughout the 1950s.

In 1958, the American Honda Company was founded and one year later, Honda introduced its first model in the United States, the 1959 Honda C100 Super Cub. The Honda Cub holds the title of being the best-selling vehicle in history, with around 50 million units sold around the world.

Honda Racing Corporation (HRC) was formed in 1954. The company combines participation in motorcycle races throughout the world with the development of high potential racing machines. Its racing activities are an important source for the creation of leading edge technologies used in the development of Honda motorcycles. HRC also contributes to the advancement of motorcycle sports through a range of activities that include sales of production racing motorcycles, support for satellite teams, and rider education programs.

Soichiro Honda, being a race driver himself, could not stay out of international motorsport. In 1959, Honda entered five motorcycles into the Isle of Man TT race, at that time the most prestigious motorcycle race in the world. While always having powerful engines, it took until 1961 for Honda to tune their chassis well enough to allow Mike Hailwood to claim their first Grand Prix victories in the 125 and 250 cc classes. Hailwood would later pick up their first senior TT wins in 1966 and 1967. Honda's race bikes were known for their "sleek & stylish design"

1949 Honda Dream D.

and exotic engine configurations, such as the 5-cylinder, 22,000 rpm, 125 cc bike and their 6-cylinder 250 cc and 380 cc bikes.

1979 saw Honda return to Grand Prix motorcycle racing with their exotic, monocoque-framed, four-stroke NR500. The NR500 featured elongated cylinders each with 8 valves and with connecting rods in pairs, in an attempt to comply with the FIM rules which limited engines to four cylinders. Honda engineered the elongated cylinders in an effort to provide the valve area of an 8-cylinder engine, hoping their four-stroke bike would be able to compete against the now dominant two-stroke racers. Unfortunately, it seemed Honda tried to accomplish too much at one time and the experiment failed. For the 1982 season, Honda debuted their first two stroke race bike, the NS500 and in 1983, Honda won their first 500 cc Grand Prix World Championship with Freddie Spencer. Since then, Honda has become a dominant marque in motorcycle Grand Prix.

In motocross, Honda has claimed 24 motocross world championships.

During the 1960s, when it was a small manufacturer, Honda broke out of the Japanese motorcycle market and began exporting to the US. Taking Honda's story as an archetype of the smaller manufacturer entering a new market already occupied by highly dominant competitors, the story of their market entry, and their subsequent huge success in the US and around the world, has been the subject of some academic controversy. Competing explanations have been advanced to explain Honda's strategy and the reasons for their success.

The first of these explanations was put forward when, in 1975, Boston Consulting Group (BCG) was commissioned by the UK government to write a report explaining why and how the British motorcycle industry had been out-competed by its Japanese competitors. The report concluded that the Japanese firms, including Honda, had sought a very high scale of production (they had made a large number of motorbikes) in order to benefit from economies of scale and learning curve effects. It blamed the decline of the British motorcycle industry on the failure of British managers to invest enough in their businesses to profit from economies of scale and scope.

The second story is told in 1984 by Richard Pascale, who had interviewed the Honda executives responsible for the firm's entry into the US market. As opposed to the tightly focused strategy of low cost and high scale that BCG accredited to Honda, Pascale found that their entry into the US market was a story of "miscalculation, serendipity, and organizational learning" – in other words, Honda's success was due to the adaptability (and hard work) of its staff, rather than any long term strategy. For example, Honda's initial plan on entering the US was to compete in large motorcycles, around 300 cc. It was only when the team found that the scooters they were using to get themselves around their US

1949 Honda Dream D.

base of San Francisco attracted positive interest from consumers that they came up with the idea of selling the Supercub.

The most recent school of thought on Honda's strategy was put forward by Gary Hamel and C. K. Prahalad in 1989. Creating the concept of core competencies with Honda as an example, they argued that Honda's success was due to its focus on leadership in the technology of internal combustion engines. For example, the high power-to-weight ratio engines Honda produced for its racing bikes provided technology and expertise which was transferable into mopeds.

Honda's entry into the US motorcycle market during the 1960s is used as a case study for teaching introductory strategy at business schools worldwide.

Honda had painstakingly marshaled a reputation of superb quality for its made-in-Japan motorcycles. But in the '70s, consumers' attitudes had been hardened by years of disappointment and frustration in American-made products, best illustrated by General Motors' problems with the Vega, Ford's with the Pinto, and the AMF-built Harley-Davidsons of the time. Honda's first U.S. plant offered an opportunity to change consumers' minds about American quality. But the effort had to succeed. If it did, Honda could proceed with the $250-million auto plant it planned to build. If the Marysville Motorcycle Plant (MMP) failed because of quality problems, Honda's reputation, and its future, could have been damaged beyond repair.

Of course, as history shows, the Marysville experiment did everything expected of it: It upheld Honda's reputation for quality and allowed the company to expand with additional plants in this country. Although the MMP was built primarily for motorcycles, the plant was designed to allow production flexibility, including the assembly of ATVs. Ten years after the first motorcycle was assembled, the best-selling ATV in America--the FourTrax® 300--rolled off the line. Honda followed the MMP with the Marysville Auto Plant (MAP, the first Japanese auto plant in the U.S.), the Anna Engine Plant (AEP), Honda Engineering North America (EGA), the East Liberty Auto Plant (ELP), Honda Transmission Manufacturing (HTM), Honda Power Equipment (HPE), and Honda of South Carolina (HSC), the company's first exclusive ATV plant.

Several reasons account for the success of Honda's U.S. plants. Start with just the physical aspects. Both the 260,000-square-foot MMP and 280,000-square-foot HSC plants are far smaller than their counterparts in Japan. Japanese plants in particular were of little use as models, in part because of their vastly greater production capacity, and because of their hodgepodge nature; they had been built and then expanded repeatedly over the years.

1949 Honda Model C.

What the American plants needed was the strictest efficiency and, in the MMP's case, flexibility to assemble different models. As Takao Shirokawa, one of the MMP's original team members, said, "The plant was designed to minimize traffic between adjacent departments, to try to minimize space, and try to maintain efficient logistics inside the plant. We tried to make the most efficient, but small, motorcycle plant. Profitability was the key to this plant. So [the question was] how to minimize the cost of assembly. We tried to pursue efficiency."

One crucial element of efficiency and quality control at HAM is the use of Honda's own assembly and production machinery. Tour the MMP and the HSC ATV plant, as well as Honda's other facilities, and you'll see the usual array of Japanese die-cast machines and American tubing-benders. But most of the high-tech precision equipment was designed and built by Honda Engineering. Welding equipment used for ATV frames, as well as stamping dies, injection molds and other machines, all bear the Honda Engineering stamp.

Building machines to make machines also provides Honda with a rapid response time.

In a competitive market, the ability to respond quickly to customer needs has given Honda an invaluable marketing edge. At Honda Engineering-including Honda Engineering of North America--engineers work closely and early in the design process with those in R&D, saving time in creating new jigs, fixtures, stamps, and dies. The arrangement also allows better, more rapid maintenance, as well as kaizen--the Japanese word for improvement--of equipment and processes.

Physically, there's little else about the MMP and HSC that break new ground. The processes and the people make the greatest difference. Take, for example, the sophisticated powder-coat paint process used for FourTrax frames at both the MMP and HSC. This efficient, high-quality, low-emissions technology received the Ohio governor's award for Outstanding Achievement in Pollution Prevention in 1998.

Apart from such purely physical aspects, the attitude the region's inhabitants bring to their work is also crucial. Before construction began on the MMP, Honda officials visited several Midwest manufacturing plants, and were convinced the people there owned a work ethic similar to that of their Japanese counterparts.

What's more, Honda treats its associates in ways almost unheard of at other plants. For instance, open communication allows associates, as Honda's employees are called, to make assembly techniques more efficient. And every associate is treated equally, down to the seemingly minor detail of all employees sharing a single common lunch room, rather than being segregated into

1958 Honda C100 Super Cub.

labor and management dining areas.

Honda also offers a single pay scale for all assembly associates, which facilitates movement from department to department. Such movement allows cross-training of associates, the development of new skills and a clearer vision of the entire assembly process that has led to many associate-driven improvements over the years. Keeping associates involved and satisfied is a key point in maintaining Honda's high level of quality.

Honda also ensures quality by working closely with its suppliers. At first that was difficult at the MMP because American suppliers were unwilling to supply Honda with parts because of the low production numbers (150 Gold Wings a day in 1981-1983 vs. 1000 units a day at one of Honda's Japanese plants). Suppliers were also unaccustomed to the level of quality Honda demanded, and were initially unwilling to make the investments necessary to provide such quality at such low volumes.

Honda persisted, though, offering training, advice, and even furnishing equipment--an unheard of relationship between client and vendor at the time, and still rare in this country. Yet most suppliers appreciated Honda's involvement, because it allowed them to improve their own quality, and so expand their business. These days, Honda of America Manufacturing (HAM) spends more than $6.4 billion on goods and services from some 450 vendors and suppliers.

Perhaps the most important way Honda ensures the highest quality in its ATVs and motorcycles is by guaranteeing and monitoring quality every step of the way. "It starts with very high-grade materials," says one company spokesman, "continues with quality suppliers, and then, in a phrase we use, quality in the process. That means we have standards, measurements, repeatable and monitorable, whether it's for a weld, or for material specs for frame material, or the quality of the injection-molded plastic material, or how long it stays in the mold--all those things are very quantifiable for high quality and durability.

"Finally, the ultimate standard we apply is customer satisfaction, which has to do with the customers' expectations and how they use the product. If they put a premium on durability and longevity and performance, we have to think up what materials and processes will ensure those things that will satisfy the customer for many years of use."

At the outset, with the construction of the Marysville

1962 Honda CB77 Super Hawk.

Motorcycle Plant, Honda took an enormous gamble. But Honda bet that the quality of its U.S.-assembled motorcycles and ATVs would lead to success, and so it ensured quality at every level, with people, processes, material, and machinery.

That commitment to quality led to the building of seven more facilities in this country over the following two decades, for a total of eight, with a total work force of more than 20,000 associates, and a capital investment of $4 billion in Honda's North American Manufacturing and R&D operations.

Honda built its reputation on that most enduring value--quality-from the very beginning, and never relented. Every motorcycle and ATV that rolls out of the MMP and HSC is a testament to that lifelong commitment.

1989 Honda NSR500.

HONDA GOLD WING

The first production model GL1000 came out in 1975, and was in production until 1979. The bike was listed as a touring bike, but it came as a bare bike. A large market developed offering fairings and luggage, the most popular being the Windjammer series by Vetter. With only minor differences for different markets this bike remained virtually unchanged for 1975-1977 production run.

In 1981, production of the Goldwing was moved from Japan to Ohio. This move brought manufacture of the motorcycle to its largest market and allowed Honda to market the machine as being made in America.

In 1982, the "Aspencade" was introduced. This was an Interstate model, with more options. AM/FM Radio, CB Radio, floorboards, and chrome were all standard on the Aspencade (these were options on the Interstate).

In 1984, the GL1200 was released, and was an immediate hit. This time the engine was totally new, and the size had grown to 1182cc. The frame was larger, and stiffened for a smoother ride. In the Interstate and Aspencade models the fairing was integrated into the main body, eliminating the appearance that they were "added on". Now the Touring models truly appeared to have been created that way.

1984 however was the last year of the "Standard" model. Over the preceding years, sales of the Standard had declined in favor of the Interstate and Aspencade models. This also lead to the decline of after-market manufacturers like Vetter.

1988 brought the most changes ever to the Goldwing. The GL1500 completely outclassed all competition to such a degree that Suzuki, Kawasaki and Yamaha eventually abandoned the market segment. The biggest difference was that the flat-4 engine was replaced with a 1520cc flat-6 engine, though it was still carbureted, Honda did introduce solid state digital ignition. This both increased power, and reduced noise. Honda also enclosed the entire motorcycle in plastic, giving it a seamless appearance.

One major innovation was the addition of a "reverse gear". Because of the size and weight, it was felt that some people would have problems backing it up.

In 2001, the first new model in 13 years was revealed. The security was so tight that nothing about it was known until it was first displayed to the public.

In September 2005, Honda announced the world's first production motorcycle airbag system scheduled for availability in late spring of 2006 in the US.

In 2004, Honda released a "Limited Edition" model, the Valkyrie Rune, complete with 1832cc engine and unique styling.

HONDA GOLD WING

PRODUCTION	1975 - present
MAX SPEED	-
ENGINE TYPE	Horizontally opposed six-cylinder
BORE/STROKE	74.0mm x 71.0mm
CAPACITY	1832cc
COMPRESSION RATIO	9.8:1
TRANSMISSION	5-speed
FINAL DRIVE	Shaft
VALVE TRAIN	SOHC; 2 Valve
CARBURETION	Programmed fuel-injection
FRONT TIRE	130/70R-18
REAR TIRE	180/60R-16
DRY WEIGHT	402 - 420 kg
WHEELBASE	1,689 mm
SEAT HEIGHT	739 mm

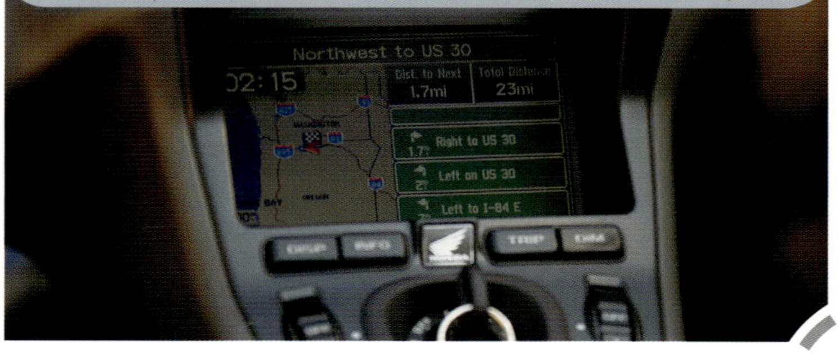

HONDA ST1300

The Honda ST1300 was introduced by Honda in 2002 and marketed in the US as a Sport Touring motorcycle. The same bike is known as the Pan-European in Europe, where it is marketed as a Touring motorcycle.

Superseding the ST1100, the bike features a standard riding posture, a liquid-cooled V4 engine and a fully-faired body with standard hard panniers.

During the 2000 bike show season, Honda began showing a prototype sport tourer called the X-Wing, which featured a 1500 cc V6 engine, single-sided front and rear suspension and an automatic transmission. Speculation in the press that the X-Wing was the ST1100's replacement was partially confirmed when Honda introduced an all-new ST1300 Pan-European in Europe and Australia for the 2002 model year. For the U.S. market, the new bike would be imported in limited numbers (about 500 per year) starting in 2003 as the ST1300.

The ST1300 incorporates many of the X-Wing's lines but none of its running gear. Power comes from a lower-slung 1261 cc V4 engine mounted as a stressed member in a lighter aluminum frame. A major difference from the ST1100 is the use of balance shafts for smoothness, allowing the engine to be directly mounted to the frame. The revised engine layout and a split fuel tank shift some of the weight downward, making the ST1300 less top-heavy than its predecessor. The rear wheel is driven through a cassette-type five-speed transmission and shaft drive.

Honda's ABS linked brake package is an option on the ST1300 in the United States, but is standard on the ST1300 Pan-European. Unlike the ST1100, the ST1300 does not include a traction control system. In 2002 and 2003, models with ABS included an electrically-adjustable windscreen, which became standard equipment on all bikes in 2004. A long list of minor differences improved upon the ST1100's comfort, handling and performance.

HONDA ST1300

PRODUCTION	2002 - present
MAX SPEED	-
ENGINE TYPE	longitudinally mounted 90° V-4
BORE/STROKE	78mm x 66mm
CAPACITY	1261cc
COMPRESSION RATIO	10.8:1
TRANSMISSION	5-speed
FINAL DRIVE	Shaft
VALVE TRAIN	DOHC; 4 Valve
CARBURETION	Programmed fuel-injection
FRONT TIRE	120/70ZR-18
REAR TIRE	170/60ZR-17
DRY WEIGHT	402 - 420 kg
WHEELBASE	1,491 mm
SEAT HEIGHT	790 mm

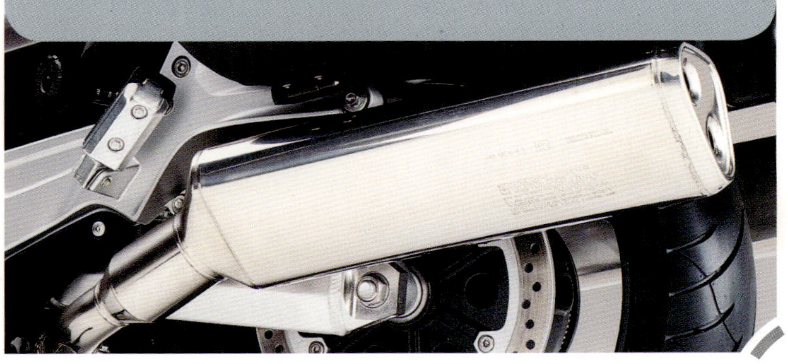

HONDA VTX

The Honda VTX series is a line of Honda V-twin cruiser motorcycles inspired by the Zodia concept shown at the 1995 Tokyo Motor Show. The Honda VTX 1800 was introduced in 2002 , and a smaller 1300 cc version was introduced in 2003. At the time this bike was introduced the Honda VTX 1800 engine was the largest displacement production V-Twin engine in the world, with the largest pistons ever produced for a car or a motorcycle. This title didn't last long however as the displacement race still continues today. There are several differently styled models available for each displacement class.

The VTX series has a muscular looking body, it is quite long and low to the ground featuring significant rake and trail. The 1800 puts out 120 ft·lbf (160 N·m) of torque at only 3500 rpm and 106 bhp @ 5000 rpm, making the VTX series one of the more powerful production V-twin motorcycles of its time.

On the VTX 1800, Honda updated its linked braking feature, instead of having the usual separate hand and foot brakes, the hand brake operates two-thirds of the front pistons while the foot operates the other third in front and all the rear via a proportioning valve. The throttle is updated with Honda's new Programmed Fuel Injection (PGM-FI) system, using a vacuum sensor within the first ten percent of opening to more precisely meter torque, until the standard throttle position sensor takes over.

Power is transmitted to the rear wheel via a shaft drive. The VTX 1300 uses standard unlinked brakes, with a single large front disk. The 1300 also uses a carburetor unlike the fuel injected 1800, and therefore is much simpler in terms of design. Although the engine is similar to the 1800, it is not just a smaller bore/stroke but a different design.

HONDA VTX 1300C	
PRODUCTION	2003 - present
MAX SPEED	-
ENGINE TYPE	52° V-twin
BORE/STROKE	89.5mm x 104.3mm
CAPACITY	1312cc
COMPRESSION RATIO	9.2:1
TRANSMISSION	5-speed
FINAL DRIVE	Shaft
VALVE TRAIN	SOHC; 3 Valve
CARBURETION	Single 38mm constant-velocity
FRONT TIRE	110/90-19 M/C
REAR TIRE	170/80-15 M/C
DRY WEIGHT	307 kg
WHEELBASE	1,663 mm
SEAT HEIGHT	697 mm

HONDA SHADOW

The Honda Shadow refers to a family of motorcyles made by Honda since 1983. The Honda Shadow is a cruiser-type motorcycle, meaning it has an upright riding position and wide handle bars. This makes it more comfortable than a sportsbike, although the lack of fairings means that the rider is exposed more to the wind. The Shadow cruisers come in many flavors, including the Spirit, Aero, Sabre, VLX, and American Classic Edition. Engine sizes range from 125cc to 1100cc. A characteristic engine for the Shadow motorcycle is a Honda VT1100 liquid-cooled, 45-degree V-twin with shaft drive.

The Shadow is available in several sizes, ranging from the leaner 125cc model to a much larger 1100cc (as the 1300cc-to-1800cc Honda VTX Series does not share the Shadow name). All Shadow cruisers are equipped with V-twin engines. In spite of this, the top speed of a 2001 Shadow 125cc, for example, is a remarkable 80 miles per hour.

The series was originally introduced as a replacement for the Honda Rebel series, although the Rebel 250cc is still produced and nearly as popular as some Shadow models.

This line of Shadows was Honda's second V-twin, a cruiser. The first being the standard Honda CX series (which in Australia was confusingly labelled the "CX500 Shadow.")

In the early 1980's there was no Shadow, although there were similar-style bikes such as the Magna, Sabre, the 1984 Nighthawk (1985 in Canada), and the Rebel. The Magna is still available as one of Honda's "other" cruisers. In 1983 the "Shadow" was born. Both the "VT750C Shadow," which incorporated the first hydraulic valve adjuster, and a "VT500C Shadow" could be found in showrooms (as noted in the Honda Motorcycle Identification Guide published by the American Honda Motor Co., Inc., in 1988).

HONDA SHADOW SPIRIT

PRODUCTION	1983 - present
MAX SPEED	-
ENGINE TYPE	45° V-twin
BORE/STROKE	87.5mm x 91.4mm
CAPACITY	1099cc
COMPRESSION RATIO	8.0:1
TRANSMISSION	5-speed
FINAL DRIVE	Shaft
VALVE TRAIN	SOHC; 3 Valve
CARBURETION	Two 36mm CV
FRONT TIRE	110/90-19
REAR TIRE	170/80-15
DRY WEIGHT	307 kg
WHEELBASE	1,651 mm
SEAT HEIGHT	729 mm

HONDA RUNE

The Honda Valkyrie is a motorcycle that was manufactured by Honda, from model years 1997–2003. The Valkyrie engine was a 1520cc horizontally-opposed six cylinder liquid cooled boxer engine transplanted from Honda's Goldwing model. This was unusual since most "cruiser" style motorcycles were based on a V-twin engine design similar to the engine of a Harley-Davidson. In its transplant from the Goldwing, the most notable engine changes were the camshaft and the change to six individual 28mm carburetors, one for each cylinder. These changes were made to increase power and torque. These changes also gave the engine a little more character by giving it a unique sound.

On introduction in 1997, a naked Standard and a Tourer model were offered. The Tourer included a windshield and lockable hard saddlebags.

In 1999, the Interstate model was added to the lineup, which included a fork-mounted fairing along with a trunk at the rear of the motorcycle.

As sales eventually dwindled, the Tourer model was dropped in 2000 and the and Interstate was dropped in 2001, leaving only the Standard model remaining. 2003 saw the Standard offered only in black and was the last year of the original Valkyrie.

Honda introduced a limited edition model in 2004 named the Valkyrie Rune with an 1800cc engine. It was a major departure from the original Valkyrie in styling, purpose and price ($25K–$26K).

HONDA RUNE

PRODUCTION	2004
MAX SPEED	-
ENGINE TYPE	horizontally opposed six-cylinder
BORE/STROKE	74.0mm x 71.0mm
CAPACITY	1832cc
COMPRESSION RATIO	9.8:1
TRANSMISSION	5-speed
FINAL DRIVE	Shaft
VALVE TRAIN	SOHC; 2 Valve
CARBURETION	PGM-FI with automatic choke
FRONT TIRE	150/60R-18
REAR TIRE	180/55R-17
DRY WEIGHT	368 kg
WHEELBASE	1,750 mm
SEAT HEIGHT	691 mm

HONDA CBR1000RR

The Honda CBR1000RR was developed by the same team that was behind the Honda RC211V race bike for the MotoGP series. Many of the new technologies introduced in the Honda CBR600RR, a direct decedent of the RC211V, were used in the new CBR1000RR such as a lengthy swingarm, Unit Pro-Link rear suspension, and Dual Stage Fuel Injection System (DSFI).

An all new CBR1000RR was introduced at the Paris International Motorcycle Show on 28 September, 2007 for the 2008 model year. The CBR1000RR is powered by an all new 999 cc (60.9 cu in) inline-four engine with a redline of 13,000 rpm. It features titanium valves and an enlarged bore with a corresponding reduced stroke. The engine has a completely new cyinder block, head configuration, and crankcase with lighter pistons. A new ECU delivers two separate revised maps sending the fuel and air mixture to be squeezed tight by the 12.3:1 compression ratio. Ram air is fed to an enlarged air box through two revised front scoops located under the headlamps. Honda claims power output to be at least 178 hp (133 kW) beginning at 12,000 rpm.

Honda made a very focused effort to reduce and centralize overall weight. A lighter, narrower die cast frame was formed using a new technique which Honda claims allows for very thin wall construction and only four castings to be welded together. Almost every part of the new bike was reengineered to reduce weight including the sidestand, front brake hoses, brake rotors, battery, and wheels.

In order to improve stability under deceleration, a slipper clutch is now available with a unique center-cam-assist mechanism. The Honda Electronic Steering Damper (HESD) has been revised this year as well. Another significant change is the exhaust system which is no longer a center-up underseat design. The CBR1000RR now features a side slung exhaust in order to increase mass centralization and compactness while mimicking a Moto GP style.

HONDA CBR1000RR

PRODUCTION	2004 - present
MAX SPEED	-
ENGINE TYPE	liquid-cooled inline four-cylinder
BORE/STROKE	76mm x 55.1mm
CAPACITY	999cc
COMPRESSION RATIO	12.3:1
TRANSMISSION	6-speed
FINAL DRIVE	O-ring chain
VALVE TRAIN	DOHC; 4 Valve
CARBURETION	DSFI
FRONT TIRE	120/70ZR-18
REAR TIRE	190/50ZR-17
DRY WEIGHT	179 kg
WHEELBASE	1,412 mm
SEAT HEIGHT	825 mm

HONDA CB900F

The Honda CB900F (also called the 919 in the U.S. and Hornet in Europe) is a "standard" or "naked" style motorcycle based on a sport bike engine but with a more upright seating position and revised engine and gearing, providing performance and comfort between a typical sport bike and a cruiser. It was introduced in 2000 and is still being manufactured today.

The CB900F is powered by a retuned Honda CBR900RR engine, developed by Tadao Baba, one of Honda's Large Project Leaders. The motor is a transversely mounted, liquid-cooled, fuel-injected 919 cc (56.1 cu in) in-line 4-stroke 4-cylinder DOHC engine that produces around 100 hp (75 kW). The engine utilizes cast camshafts and pistons instead of the pricier forged items. For greater midrange punch, the CB900F's camshaft profiles are milder and compression is slightly lowered. Four 36 mm (1.4 in) fuel-injection throttle bodies take the place of the CBR900RR's 38 mm (1.5 in) carburetors. Redline happens at a 9,500 rpm and the bike has a six-speed transmission.

A steel, square-tube backbone frame supports the engine as a stressed member. In front, a cartridge fork (adjustable beginning in 2004) guides the wheel, while a single shock, adjustable only for preload,(& rebound dampening beginning in 2004) connects with the aluminum swingarm and carries the weight in back. Its brakes are dual-disc in the front and single-disc in the rear.

Instrumentation consists of an analog speedometer and tachometer and basic indicator lamps, incorporated under a tinted window, and a single tripometer.

While the CB900F comes sans centerstand, one is offered as an accessory for 49-state models, although it can be fitted to a California model. The bike's rake is 25°, trail is 98 mm (3.9 in), wheelbase is 1460 mm (57.5 in), and seat height is 800 mm (31.5 in). It has a tested dry weight (minus fuel only) of 455 lb (206 kg) and a tested wet weight of 485 lb (220 kg).

Honda has had a CB900 model since 1980 in North America and an even earlier CB900F model in Europe. A 599 cc (36.5 cu in) carburetted version called the CB600F exists. The US models can not use the Hornet moniker due to the name being trademarked by the American Motor Corporation.

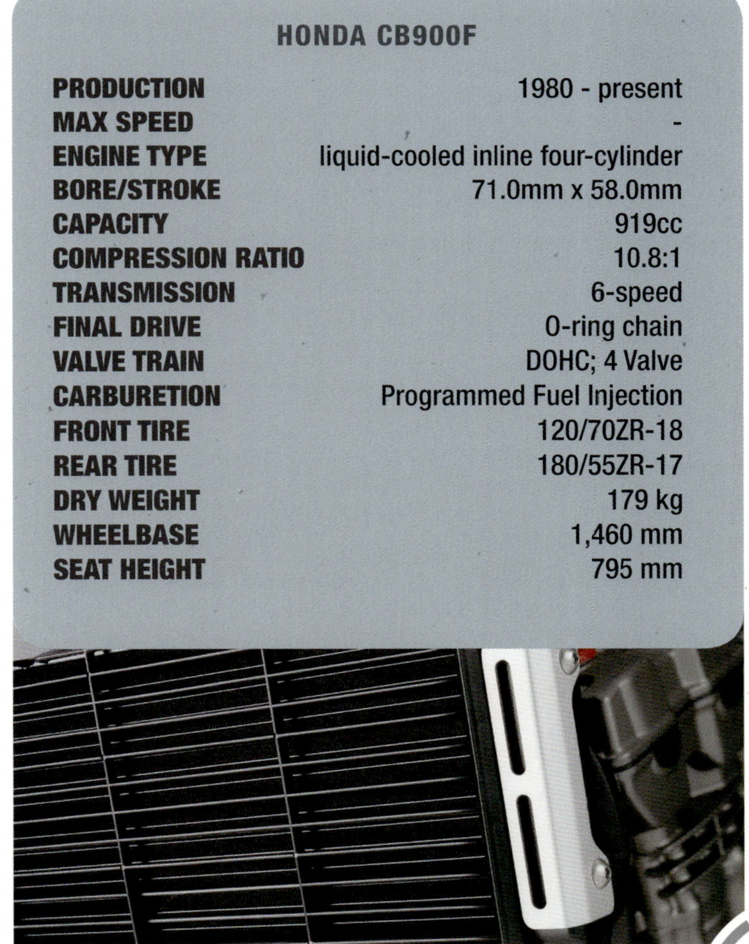

HONDA CB900F

PRODUCTION	1980 - present
MAX SPEED	-
ENGINE TYPE	liquid-cooled inline four-cylinder
BORE/STROKE	71.0mm x 58.0mm
CAPACITY	919cc
COMPRESSION RATIO	10.8:1
TRANSMISSION	6-speed
FINAL DRIVE	O-ring chain
VALVE TRAIN	DOHC; 4 Valve
CARBURETION	Programmed Fuel Injection
FRONT TIRE	120/70ZR-18
REAR TIRE	180/55ZR-17
DRY WEIGHT	179 kg
WHEELBASE	1,460 mm
SEAT HEIGHT	795 mm

HONDA INTERCEPTOR

The Honda VFR800/Interceptor is a motorcycle introduced by the Honda Motor Company in 1998. It is a successor to the VFR750F (1986-1997), which was preceded by the VF750, a machine with camshafts prone to failure. Accordingly, the VFR750F motor was stoutly engineered with gear-driven camshafts to salvage Hondas reputation and ensure such failures did not reoccur.

From its first sales in 1986, the VFR750F scored highly on many press reviews. While it was originally intending to be a sports bike, the introduction of lighter competitors (including the nearly 30 kg lighter GSX-R750) prompted Honda in 1990 to transition the VFR into a mid-sized sports-touring bike, a category of which the VFR became the de-facto benchmark. However, by 1997, Triumph and Ducati presented Honda with significant sports-touring competition, so Honda responded with a redesigned VFR800 in 1998.

The sixth generation VFR (for the first time marketed only as the "Interceptor") was introduced in 2002. For the first time it featured dual underseat exhausts, an optional anti-lock braking system (ABS) in addition to the DCBS, and optional hard luggage. It featured chain-driven cams rather than the traditional VFR gear-driven cams, and was the first motorcycle to have Honda's VTEC valve-actuation technology.

The implementation of VTEC was seen as a bid by Honda to meet tightening noise and emissions standards in Japan and abroad while improving the peak horsepower of the engine.

HONDA INTERCEPTOR

PRODUCTION	1998 - present
MAX SPEED	-
ENGINE TYPE	liquid-cooled inline four-cylinder
BORE/STROKE	72mm x 48mm
CAPACITY	781cc
COMPRESSION RATIO	11.6:1
TRANSMISSION	6-speed
FINAL DRIVE	O-ring chain
VALVE TRAIN	VTEC DOHC; 4 Valve
CARBURETION	Programmed Fuel Injection
FRONT TIRE	120/70ZR-17
REAR TIRE	180/55ZR-17
DRY WEIGHT	218.2 kg
WHEELBASE	1,458 mm
SEAT HEIGHT	805 mm

HONDA CBR600RR

The CBR600RR was developed from and inspired by the Honda RC211V MotoGP bike. The similar physical appearance of the CBR600RR and RC211V is intentional. Underneath the looks lie MotoGP technologies that were made available for the first time on a production motorcycle such as the Unit Pro-Link rear suspension and Dual Stage Fuel Injection (PGM-DSFI). Both were taken directly from Honda's MotoGP bike. While it's not uncommon for street bikes to utilize racing technology, this was the first time totally new technologies found their way to the production line the same year they made their way to the professional racing grid. The bike receives the "RR" designation for "race replica" because of its emphasis on racing characteristics such as an advanced braced swingarm, center-up exhaust system, and more aggressive riding position. The 2003 model carried over to 2004 technically unchanged, but with new color schemes.

Honda's middleweight sport bike underwent a complete and total redesign for 2007 with reducing the CBR600RR's weight as their primary focus. The result was a 20 lb (9.1 kg) reduction in dry weight over the 2006 model, from 361 lb (163.7 kg) to 341 lb (154.7 kg).

In redesigning the CBR600RR for lighter weight and increased performance, Honda's R&D engineers started with the engine. The completely new engine is the smallest and lightest powerplant in the super sport class by a significant margin - its designers having used careful positioning of all internal components to achieve significant reductions in the motor's length, width, and height, as well as reducing weight by 2 kg (4.4 lb) compared to the 2006 model's powerplant. While the new engine is smaller and lighter than its predecessor, it also offers increased performance, to the tune of a manufacture claimed 118 hp (88 kW) at 13,500 rpm. At speed, that figure is increased significantly by the highly efficient centrally-located ram-air duct in the bike's nose.

HONDA CBR600RR	
PRODUCTION	1998 - present
MAX SPEED	-
ENGINE TYPE	liquid-cooled inline four-cylinder
BORE/STROKE	72mm x 48mm
CAPACITY	781cc
COMPRESSION RATIO	11.6:1
TRANSMISSION	6-speed
FINAL DRIVE	O-ring chain
VALVE TRAIN	VTEC DOHC; 4 Valve
CARBURETION	Programmed Fuel Injection
FRONT TIRE	120/70ZR-17
REAR TIRE	180/55ZR-17
DRY WEIGHT	218.2 kg
WHEELBASE	1,458 mm
SEAT HEIGHT	805 mm

Arai
Ninja

KAWASAKI

1967 Kawasaki A1

Kawasaki's origins go back to 1876, when Shozo Kawasaki established Kawasaki Tsukiji Shipyard in Tokyo. Eighteen years later, in 1896, it was incorporated as Kawasaki Dockyard Co., Ltd.

Born in Kagoshima to a kimono merchant, Shozo Kawasaki became a tradesman at the age of 17 in Nagasaki, the only place in Japan then open to the West. He started a shipping business in Osaka at 27, which failed when his cargo ship sank during a storm. In 1869, he joined a company handling sugar from Ryukyu (currently Okinawa Prefecture), established by a Kagoshima samurai, and in 1893, researched Ryukyu sugar and sea routes to Ryukyu at the request of the Ministry of Finance. In 1894, he was appointed executive vice president of Japan Mail Steam-Powered Shipping Company, and succeeded in opening a sea route to Ryukyu and transporting sugar to mainland Japan.

Having experienced many sea accidents in his life, Kawasaki deepened his trust in Western ships because they were more spacious, stable and faster than typical Japanese ships. At the same time, he became very interested in the modern shipbuilding industry. In April 1878, supported by Masayoshi Matsukata, the Vice Minister of Finance, who was from the same province as Kawasaki, he established Kawasaki Tsukiji Shipyard on borrowed land from the government alongside the Sumidagawa River, Tsukiji Minami-Iizaka-cho (currently Tsukiji 7-chome, Chuo-ku), Tokyo, a major step forward as a shipbuilder.

Kawasaki's Aircraft Company began the development of a motorcycle engine in 1949. The development was completed in 1952 and mass production started in 1953. The engine was an air-cooled, 148cc, OHV, 4-stroke single cylinder with a maximum power of 4 PS (3.9 hp/2.9 kW) at 4,000 rpm.

In 1954 the first complete Kawasaki Motorcycle was produced under the name of Meihatsu, a subsidiary of Kawasaki Aircraft.

In 1960 Kawasaki completed construction of a factory dedicated exclusively to motorcycle production and bought Meguro Motorcycles.

The Japanese economy experienced a period of particularly high economic growth from the mid-1950s until the mid-1960s. Kawasaki grabbed this opportunity with both hands and the company's production of motorbikes surged in response. Then some time around the middle the 1960s marked the start of the Supersports era.

Kawasaki rode the wave generated by what later came to be called the "Izanagi Boom" and, in September 1966, the company proudly announced its W1, a motorcycle with a displacement of 650cc - at the time of its launch, this was the largest displacement bike on the Japanese market. In August 1967, the A1, more commonly known as the "Samurai", took center stage - this was a high performance machine with approximately 80ps per liter. It was quickly followed by a larger bore model, the A7. This was undoubtedly the era when Kawasaki laid the foundations for its current success.

The history of the W1 can be traced back to 1960 and the early K1, a motorcycle developed by the Japanese motorcycle meguro. Meguro had first started producing motorcycles back in 1909 and had modeled the K1 on the English BSA A7 as a replacement for their single cylinder Meguro Z7.

It was early days, and most Japanese motorcycle manufacturers at the time were basically building bikes copied from American and European models, particularly in the large displacement categories.

For its day, the K1 was an advanced design and showcased modern-day manufacturing techniques with its Air-Cooled , 4-stroke, Twin OHV 496cc engine mounted in a double-cradle frame.

In 1960, Meguro Works entered into a business relationship with Kawasaki Aircraft Co.,Ltd., leading to a full merger in 1963. Therefore, although the K1 was developed and produced by Meguro, selling it was left to Kawasaki Motor Sales Co., the forerunner of Kawasaki Motorcycle Co.,Ltd. At the time, the Kawasaki engineers were so deeply engaged in the development of a 4-stroke engine for small cars that they had no time to develop a new motorcycle engine. But by the end of 1962 the four-wheel project had ended and some of these car engineers transferred to Meguro Works. There were two projects that the developers had to tackle: the SG(a single-cylinder 250cc OHV) and the K1.

For both projects, ex-Meguro engineers kept working on the task of chassis development, while the SG and K1 engines were developed by the ex-Kawasaki engineers. Since the K models were still in the transition stage from Meguro to Kawasaki, there were many problems associated with technology transfers and maintenance. However, work proceeded at the same time on development of a successor to the K1 \ the new W1. At the time, sales objectives were concentrated on receiving orders for police patrol motorcycles intended for guard duties during the 1964 Tokyo Olympics. Since there was no time to develop a new engine, or even to remodel an existing one, there was no choice but to use the K1 model as it was. However, the engineers still wanted to overcome some basic design flaws in the K1 engine. Because the sales side wanted to maintain the impressive appearance and dignified look of the K1, it was decided to remodel the engine only, and in 1965 the remodeled K1 was introduced as the K2.

Kawasaki, however, was not the only manufacturer that had the presence to take advantage of those economic boom years. Honda launched its CB450 during the same time and Suzuki its T500. Both these models proved to be very popular with enthusiasts overseas.

At the time, the motorcycle riding public keenly sought larger bore machines with higher horsepower and higher top speeds and each company vied to provide what the market wanted. In the US - America was the world's biggest motorbike market - riders were particularly insistent on these new features. Unfortunately, when these demands reached the desks of the manufacturers back in Japan, they usually were accompanied by calls for lower prices. Obviously this was not going to be easy for the Japanese bike manufacturers.

Kawasaki's response was to put together a top secret plan referred to as the N100 Plan. Development got under way in earnest in July 1967 with the eventual target becoming a set of specifications that were nothing short of spectacular for the period: an engine capacity of 500cc, 60ps output (equivalent to a per liter horsepower of 120ps) and a 0-400m standing start time of 13 seconds.

Until 1968, Kawasaki was mainly involved in developing two stroke motorcycles, although the company did have a long history of developing four stroke engines. In 1937, Meguro (merged with Kawasaki in 1963) manufactured 500cc single engines, and the engineers who developed this technology moved to Kawasaki. These engineers played a major role in developing the 650cc W series motorcycles.

This experience provided Kawasaki with the basic skills to develop four stroke engines.In 1967 Kawasaki made a decision to develop a high-performance motorcycle which would far exceed the 650W1, the largest motorcycles in Japan that time.

As the United States was targeted as the main market for these high performance motorcycles, the development team was sent to the U.S. where they secretly worked out a plan for the new model.

Finally, the displacement of the new model was set at 750cc and a mock-up was completed in October 1968. However, Honda announced a new 750cc single-over-head-cam (SOHC) motorcycle at the Tokyo Motor Show held the same year. The Kawasaki management staff realized it was meaningless to come out with a similar model after Honda had already introduced theirs, so all development efforts on Kawasaki's 750cc model were stopped.

In 1970, the Z1 (development code T103) developing project team was reunited with the best staff in all the fields joining the project. Kawasaki resumed U.S. market research in March of 1970 and collected customers' opinions from various sources such as random samplings of dealers and editors of major motorcycle magazines.

Finally, the management staff concluded there was a strong market for a high-speed, attractive motorcycle with enough power to use as a reliable touring model.

Kawasaki's answer to this market was a 1,000cc class, four stroke, four cylinder model. The main requirements for the Z1 engine were high speed, high stability, and ease of dealing with pollution problems. A four stroke unit meeting these requirements would be met by strong market appeal.

The first prototype was completed in the spring of 1971. This prototype was ridden by American test riders with minor adjustments made step by step. In the fall of that year, the final prototype was completed and after testing, the unit was approved for mass production. The first production model was completed in February 1972, and this unit was subjected to repeated severe road testing after which all parts, including even the nuts and bolts, were examined. After reworking all weak points, the first mass-production model was built in May 1972.

The 903cc displacement of the Z1 made it the largest motorcycle in Japan. Worldwide, it was larger than Italian Moto Guzzi 850 and comparable to Harley-Davidson 1000 and 1200.

The specifications called for an air-cooled four-stroke four-cylinder engine with a double-over-head-cam (DOHC) mechanism.

1970's Kawasaki Z1

The DOHC was necessary to realize overall high performance from low to high speeds. In motorcycle markets around the world, there were only one or two other samples of this type of engine, and it was the first engine for Kawasaki to adopt this advanced valve train.

In September 1972, the Z1 was introduced to the U.S. public, and sales started in November of that year. Since the development stage, Z1 was nicknamed "The New York Steak," and the Z1 was enthusiastically welcomed by markets as the "mouth watering motorcycle" when sales started. The suggested retail price was $1,900 and the initial sales plan called for 1,500 vehicles per month including the European markets.

Eleven years after the introduction of the Z1, Kawasaki's newest 900cc could again claim the title "World's Fastest." Its recorded top speed was over 240km/h and its 0-400m acceleration time was 10.976 seconds. After tireless preparation, Kawasaki introduced the new GPZ900R to the public at the 1983 Paris Salon. Its outstanding performance capabilities utterly impressed all who attended the GPZ900R world press introduction at Laguna Seca Raceway, USA in December of the same year. In January 1984,

1973 Kawasaki F11

sales of the GPZ900R began world-wide with the word "Ninja" added as a prefix to its name for the North American market. It didn't take long before the GPZ900R became the best-selling bike in the world. It won the title "Bike of the Year" in many countries.

The period from the late sixties to the seventies was the one of the most eventful in motorcycling history as the application of revolutionary technology dramatically increased motorcycle performance, multi-cylinder two-strokes with ever larger displacements, large-displacement four-strokes with DOHC engines-a veritable avalanche of new technology changed forever the face of motorcycling.

Kawasaki, one of the main instigators of this revolution, stunned the two-wheeled community early on with the release of the 500SS/MachIII, a shockingly powerful three-cylinder two-stroke whose awesome performance left an indelible mark on motorcycling history. It was followed by the mighty Z1, a big-bore four-stroke "muscle bike" delivering an asphalt ripping level of performance never before seen. Of course, other manufacturers also built a variety of large-displacement high-performance machines, fueling an escalating horsepower war the likes of which motorcycling had never seen.

A brief cessation of rivalry was called in October of 1972, due to, a world-wide oil crises. This was caused by OPECs raising the price of crude oil, and a consequent emphasis on fuel efficient, environmentally-friendly engines. Early victims were two-strokes and large-displacement four-strokes, which were seen as politically incorrect long before the term came into usage. At the time, Kawasaki was secretly developing a liquid-cooled 750cc two-stroke four, and a rotary engine, both of which had to be abandoned due to fuel consumption considerations.

In spite of all this, the Z1 continued to sell strongly, as a growing legion of hard-core fans found themselves unable to kick the high-horsepower habit. Realizing that demands for high-performance supersport bikes still existed, in 1974 development was quietly begun on a new flagship model to replace the famous Z1.

The new model would incorporate the latest technology, would have displacement befitting its flagship status; and, of course, would, offer the awesome performance that had become a Kawasaki trademark. The development concept was both simple

1978 Kawasaki Z1R

and bold: create a high-quality sport tourer incorporating the best of Kawasaki's technology, and one that would provide even more performance than the Z1.

In the new political and environmental climate, engine development was of the utmost importance. During the planning stage, a displacement of 1,200cc was called for, -1.3 times that of the Z1. Following the release of the Z1, Kawasaki was the undisputed sales leader when it came to large-displacement DOHC air-cooled, four-strokes, but the Z1's technology was no longer cutting edge. Looking for a high-impact engine design for the new era, various layouts were discussed and discarded, including V-fours and square-fours. At the suggestion of Kawasaki Motors Corp., in the United States, a V6 monster engine was mulled. The political and technological nouveau demanded a liquid-cooled four-stroke with DOHC head. After much discussion, it was decided that the new machine would be powered by an in-line six -Kawasaki's first attempt at building a monster motorcycle engine.

Stylistically, it was decided that the machine would feature a more advanced expression of the supersport styling used on the Z1. The new liquid-cooled six-cylinder engine would use some technology from the Z1, but a one-piece, plain-bearing crank would be used instead of the Z1's built-up unit. Of particular concern was crankshaft twisting. This problem was solved by strengthening the crank and increasing its journal size. The frame would also have to deal with increased engine weight and power, and many frames were concerned with reinforced steering heads and other features to deal with the expected 120 hp. The drive train for the new machines was also a question mark, with shaft drive being seriously considered for what was to be a sport touring machine.

Kawasaki was making transmissions for Isuzu at the time, and was one of the nation's leading manufacturers in this field. This technology was used to develop a shaft drive system for the new bike. Compared with a chain, the shaft requires less maintenance, and the sacrifice in power and weight was considered acceptable given the bike's sport touring orientation. Coincidentally, both Honda and Yamaha were starting to use shaft drives at the same time on the GL1000 and XS1000, respectively.

The first prototype was completed in August of 1976. It was done in a supersport style, and sported a bikini cowl. Its slim chassis was accentuated by the impressive engine with its bank of six carburettor and a six-into-two exhaust system. But during the two years of its development, market tastes had shifted from

supersport towards touring. In response, the new flagship was given a larger fuel tank and restyled as a big tourer.

In 1977, in response to Harley-Davidson's announcement of a new 1,340cc engine, the displacement of the new bike was bumped from 1,200cc to 1,286cc, and fuel tank capacity was increased to 27 litres for long-distance touring.

The final, restyled prototype was completed in March of 1978, and full-scale testing was started. To lighten the handling of what was expected be the most massive production bike until that time, considerable effort was expended in tuning the frame geometry and suspension. After building and discarding more than twenty different frames, a combination was found which delivered handling performance equal to that of the Z1. And, since this flagship model would be the successor to the Z1, thorough attention was given to every detail. To reduce vibration and mechanical noise, many tests were carried out in a soundproof room.

With final testing complete, the Z1300 made its world debut at the Koln show in September of 1978. The liquid-cooled, 1,286cc, DOHC, two-valve engine had a maximum output of 120ps@8,000 rpm and maximum torque of 11.8kg-m@6,000 rpm which, combined with dry weight of 297 kg, made it the super dreadnought of touring bikes.

1989 Kawasaki ZX600

Press introduction for the Z1300 were held in November 1978 on the island of Malta for the European press and in Death Valley, California, for the American press. All were unanimous in their praise of both the styling and performance of this ground-breaking new machine. In spite of a wet weight exceeding 300 kg, the Z1300 had no trouble launching the impressive bike into motion -all of which was dutifully reported to motorcycling fans around the world. In Germany, site of Koln show, the Z1300 soon became a collector's item.

The Z1300 was produced from 1978 to 1983, when it was given a digital fuel injection("D.F.I.") system and redesignated as the ZG1300 (although the generic model name continued as Z1300). Although this system was installed primarily to improve fuel consumption, even with no other changes to the motor, maximum power jumped to 130ps@8,000 rpm and maximum torque to 11.8kg-m@7,000 rpm.

KAWASAKI NINJA ZX-10R

The Kawasaki Ninja ZX-10R is Kawasaki's follow-up to the ZX-9R sport bike. It was originally released in 2004 with minor revisions in 2005, it combines ultra-narrow chassis, low weight, radial brakes, and strong engine make it a very competitive package for its Japanese rivals: Suzuki's GSX-R 1000, Honda's CBR1000RR and Yamaha Motor Company's R1. In 2004 and 2005 the ZX-10R won Best Superbike from Cycle World magazine and the prestigious international Masterbike competition. It is known to be the most 'hard edged' or race oriented out of the Japanese inline four cylinder motorcycle, with relatively quick steering and a ferocious powerband.

A complete overhaul to the ZX-10R in 2006 sees the most comprehensive round of changes since the bike was introduced. Kawasaki engineers utilized a "stack" design for the extraordinarily compact, liquid-cooled, 998 cc inline four-cylinder engine. The crank axis, input shaft and output shaft of the "Ninja" ZX-10R engine are positioned in a triangular layout to reduce engine length, while the high-speed generator is placed behind the cylinder bank to reduce engine width. With a bore and stroke of 76 x 55 mm, the ZX-10R engine's one-piece cylinder and crankcase assembly reduces weight and increases rigidity. The double overhead cams are machined from chromoly steel billet for strength, four valves per cylinder improve high-rpm breathing, and the forged, lightweight pistons offer high heat resistance to further enhance the bike's power-to-weight ratio.

KAWASAKI NINJA ZX-10R

PRODUCTION	2004 - present
MAX SPEED	-
ENGINE TYPE	liquid-cooled inline four-cylinder
BORE/STROKE	76.0 x 55.0 mm
CAPACITY	998cc
COMPRESSION RATIO	12.9:1
TRANSMISSION	6-speed
FINAL DRIVE	Sealed chain
VALVE TRAIN	DOHC, 16 valve
CARBURETION	Fuel Injection
FRONT TIRE	120/70ZR-17
REAR TIRE	190/55ZR-17
DRY WEIGHT	179 kg
WHEELBASE	1,415 mm
SEAT HEIGHT	830 mm

KAWASAKI Z1000

Kawasaki introduced the kz900 motorcycle in 1973. Four years later they introduced the kz1000. The original 1973 Kawasaki Z900/Z1/Z-1 was the first Japanese motorcycle with four cylinders, dual overhead cams and 903 cubic centimeters (cc's). It was one of the most powerful motorcycles produced up until that time. And even though its quarter mile times and top speeds were grossly overestimated, actual numbers were impressive enough to earn the bike the nickname "The King." In 1977 a Z1000 ridden by Reg Pridmore became the first Japanese bike to win an AMA Superbike national when it took the victory at Pocono Raceway. In 2003 Kawasaki introduced a completely revamped 30 year anniversary edition of the Z1000. It used a modified motor from the Kawasaki ZX-9R, and was bored out by 2.2 mm resulting in bigger displacement, more low-RPM torque, and only a slight power loss of 4bhp from the original ZX9 lump. In 2004, Kawasaki released the Z1000's smaller brother, the Z750. In 2007, Kawasaki re-defined the naked class with an improved version of the previous model as far as mechanics go, and Kawasaki went wild with this one stylistically. The Z1000 is also known as the "Z," "Zed," and "Z1k."

KAWASAKI Z1000

PRODUCTION	2003 - present
MAX SPEED	-
ENGINE TYPE	liquid-cooled inline four-cylinder
BORE/STROKE	77.20 x 50.9 mm
CAPACITY	953cc
COMPRESSION RATIO	11.2:1
TRANSMISSION	6-speed
FINAL DRIVE	X-Ring Chain
VALVE TRAIN	DOHC, 4 valve
CARBURETION	Fuel Injection
FRONT TIRE	120/70ZR-17
REAR TIRE	190/50ZR-17
DRY WEIGHT	205 kg
WHEELBASE	1,445 mm
SEAT HEIGHT	820 mm

KAWASAKI Z750

The Z750 motorcycle is Kawasaki's model in the budget class of naked and half faired bikes. It is a smaller version of the Kawasaki Z1000.

The Kawasaki Z750 was launched in 2004, after its bigger brother, the Z1000 in 2003. Kawasaki kept it simple, using the same engine block and sleeving it down from 1000cc to 750cc, cheaper front suspension and using a conventional exhaust, making it a cut-down version of Z1000. Like the Z1000, which is considered a modern version of the Kawasaki Z900/Z1, the Z750 can be considered a modern take on the Kawasaki Z750RS Z2. A remodeled version of both the Z750 and the Z1000 was released in 2007, with changes both stylistically and mechanically.

In 2005, Kawasaki launched the Z750S version which is more tourer, and less streetfighter. This version has a single long seat instead of the two-part seat on the Z750 similar to the Kawasaki ZXR600R, half fairing for wind protection, and excludes the rear tire "hugger" as found on the unfaired Z750. The 'S' version also uses an analog speedometer and tachometer instead of the digital instrument cluster taken from Kawasaki's supersport ZX-R models of which some riders complain that the digital LCD tachometer was harder to read while riding. Other differences inclue a slightly lower seat, grab rails and ZX10 style rear brake lights.

KAWASAKI Z750

PRODUCTION	2004 - present
MAX SPEED	-
ENGINE TYPE	liquid-cooled inline four-cylinder
BORE/STROKE	68.4 x 50.9 mm
CAPACITY	748cc
COMPRESSION RATIO	11.3:1
TRANSMISSION	6-speed
FINAL DRIVE	Sealed chain
VALVE TRAIN	DOHC, 16 valve
CARBURETION	Fuel Injection
FRONT TIRE	120/70ZR-17
REAR TIRE	180/55ZR-17
DRY WEIGHT	195 kg
WHEELBASE	1,425 mm
SEAT HEIGHT	815 mm

KAWASAKI ER-6F

The Kawasaki Ninja 650R is the U.S.-market, faired version of the motorcycle known as the ER-6n elsewhere in the world, was introduced in 2006. The 650R is a middleweight twin motorcycle, designed for normal use on paved roads. It has modern styling and features, with low-seating ergonomics, a low center of gravity, and respectable, manageable power output, and has proven to be enough bike to keep seasoned riders entertained. The design of the bike is meant to maximize rider comfort and integration, and be aesthetically equivalent to larger, more powerful superbikes.

The 650R/faired ER-6, known as the ER-6f overseas, was introduced to the market in 2006 by Kawasaki Motorcycles. The unfaired ER-6n is not sold in the U.S. The 2008 650R has an MSRP of $6,499 USD, $200 more than the 2006 model. The motorcycle fits above the Ninja 250R & Ninja 500R models which were already present in Kawasaki's sportbike lineup, which includes the famous Ninja ZX models. 2006 U.S. models came in two color options; black upper with silver bottom, and silver upper with black bottom. The 2008 models are available in Sunbeam Red, Lime Green, and Metallic Diablo Black, in single-tone treatment.

In Europe the bike is sold as the ER-6f or faired version of the ER-6, the "naked" roadster version is sold as the ER-6n. The ER-6f differed slightly from the Ninja 650R as it featured the passenger handlebars as standard and a full single colored fairing. The ER-6n comes in 4 color options, orange, yellow or black with a champagne frame, and silver with a red frame. In addition the option of ABS brakes were made available for both the ER-6n and ER-6f. There is also a derivative of the ER-6 called the Versys which utilises many of the component parts. In many European countries the ER-6n naked version has proven considerably more popular than the 6f.

KAWASAKI ER-6F	
PRODUCTION	2004 - present
MAX SPEED	-
ENGINE TYPE	liquid-cooled Parallel Twin
BORE/STROKE	83 x 60 mm
CAPACITY	649cc
COMPRESSION RATIO	11.3:1
TRANSMISSION	6-speed
FINAL DRIVE	O-Ring Chain
VALVE TRAIN	DOHC, 8 valve
CARBURETION	Fuel Injection
FRONT TIRE	120/70ZR-17
REAR TIRE	160/60ZR-17
DRY WEIGHT	178 kg
WHEELBASE	1,410 mm
SEAT HEIGHT	790 mm

KAWASAKI NINJA 250R

Kawasaki has marketed the Ninja 250R since 1986 as an entry-level sport motorcycle intended for normal use on paved roads. The bike is marketed in the US as the Ninja 250R, as the ZX250 in the UK -- and as the GPX250 elsewhere.

According to a 2008 MotorcycleUSA.com article, the bike has been Kawasaki's best-selling bike, "experiencing steady double-digit sales growth year after year."

In many respects, including ergonomics, chassis design, engine placement within the frame, the Ninja 250R straddles standard and sport classes. Likewise, the bike's riding position falls between standard and sport. Capable of running the 1/4 mile in 14.6 seconds at 88mph, the bike's features include bungee hooks, centerstand, a tachometer, and front and rear disc brakes.

The bike has been heavily updated for 2008 with a new full fairing and 17" wheels. As of 2007, the 250R has been produced in Thailand.

For the 2008 model year, the Ninja 250 increased in price by $500 US, to $3,499, and saw the first major design changes in some two decades.

The most significant changes are the loss of the center stand, an overall more modern look, and increasing the wheel size to 17",. The front brake was also enlarged, the front suspension was increased in diameter and the dual exhaust was changed to a 2 into 1 layout. The addition of a fuel gauge, a rarity in motorcycles, and the choice to make the speedometer more prominent than the tachometer imply an emphasis on attracting new riders rather than enthusiasts.

The engine and drive train are 70% new, retaining only 30% of the older models parts, according to Kawasaki's marketing literature. Compression and maximum torque have been lowered for 20% better midrange performance. These changes add up to a signifigant improvement over the previous model.

KAWASAKI NINJA 250R

PRODUCTION	1986 - present
MAX SPEED	-
ENGINE TYPE	liquid-cooled Parallel Twin
BORE/STROKE	62.0 x 41.2 mm
CAPACITY	249cc
COMPRESSION RATIO	11.6:1
TRANSMISSION	6-speed
FINAL DRIVE	Sealed chain
VALVE TRAIN	DOHC, 8 valve
CARBURETION	Fuel Injection
FRONT TIRE	110/70ZR-17
REAR TIRE	130/70ZR-17
DRY WEIGHT	178 kg
WHEELBASE	1,400 mm
SEAT HEIGHT	775 mm

KAWASAKI 1400GTR

In September 2006, Kawasaki announced a new generation Concours, know as the Concours 14 in the USA, or 1400GTR in other markets. Introduced in September 2007 as a 2008 model, the new bike is based on the ZX-14 platform with features similar to the original Concours – an inline-4 engine, luggage, shaft drive and a full fairing.

In addition to optional ABS, the new bike offers an electrically controlled screen, an innovative pass-key system, and a sophisticated rear suspension-drive system known as Tetralever – not unlike BMW's Paralever and Moto Guzzi's CARC rear suspensions – to handle the conflicting drive and suspension forces (known as shaft effect) typical when shaft-driven motorcycles carry powerful engines.

KAWASAKI 1400GTR

PRODUCTION	2006 - present
MAX SPEED	-
ENGINE TYPE	liquid-cooled in-line
BORE/STROKE	84.0 x 61.0 mm
CAPACITY	1,352cc
COMPRESSION RATIO	10.7:1
TRANSMISSION	6-speed
FINAL DRIVE	Shaft
VALVE TRAIN	DOHC, 16 valve
CARBURETION	Fuel Injection
FRONT TIRE	120/70ZR-17
REAR TIRE	190/50ZR-17
DRY WEIGHT	279 kg
WHEELBASE	1,520 mm
SEAT HEIGHT	815 mm

KAWASAKI VN900 CLASSIC

The Kawasaki Vulcan 900 Classic motorcycle (Model VN900B) is a mid-sized motorcycle cruiser made by Kawasaki Heavy Industries, first introduced in 2006 and is still in production today. The cycle follows the formula of a smaller yet capable engine fitted into a one-size up frame, a popular combination also in use by Honda, Suzuki, and Yamaha in their respective cruiser lines, and is currently possesses the largest displacement in the middleweight cruiser class.

The VN900B is a boulevard-style cruiser, similiar in appearance to the Harley-Davidson Softail Deluxe or the Fat Boy. It is powered by a liquid cooled 903cc / 55.1 cu in V-Twin engine, rated at 60.8 lbs-ft of torque @ 3,700 rpm, with a five speed transmission. Overall, it measures 97.0 inches in length, has a wheelbase of 64.8 inches, and possesses a seatheight off the ground of 26.8 inches. Dry weight is 557 lbs.

The VN900D is the touring edition of the basic VN900B. The major additions made to accommodate this new role include factory installed windshield, saddlebags, and a backrest. Additionally, a studded seat replaces the standard unit.

Also available is the VN900C, which is regarded as the more aggressively styled sister of the VN900B. Major changes are cast alloy wheels (solid on the rear, and spoked for the front), a much smaller diameter 21 in 80/90 front wheel, redesigned fenders, forward controls with pegs (as opposed to floorboards), a smaller headlight, and drag-style bars.

KAWASAKI VN900 CLASSIC	
PRODUCTION	2006 - present
MAX SPEED	-
ENGINE TYPE	liquid-cooled V twin
BORE/STROKE	88.0 x 74.2 mm
CAPACITY	903cc
COMPRESSION RATIO	9.5:1
TRANSMISSION	6-speed
FINAL DRIVE	Shaft
VALVE TRAIN	SOHC, 8 valve
CARBURETION	Fuel Injection
FRONT TIRE	130/90ZR-17
REAR TIRE	180/70ZR-17
DRY WEIGHT	253 kg
WHEELBASE	1,650 mm
SEAT HEIGHT	680 mm

KAWASAKI VERSYS

Kawasaki introduced the Versys, a middleweight motorcycle with standard riding posture, to the European and Canadian markets at the end of 2006, and to the U. S. market in 2007.

The name Versys is a portmanteau of the words "versatile system," suggesting a system of riding attributes intended to offer versatility.

Based on the Ninja 650R, the bike's 650cc liquid cooled 4 stroke parallel twin engine has been retuned for more bottom-end and mid-range torque. This is achieved with revised inlet and exhaust cams with shorter valve duration. This will move peak torque lower down the rev range providing a better throttle response at low revs. In addition a balance tube has been added between the exhaust headers to smooth out power delivery.

This lowering of peak torque in the rev range may give a trade off against peak power output. The current ER-6 produces a claimed 71bhp at 8500rpm. In comparison the Versys peak power output will be 64bhp at 8000rpm. Peak torque will be 61 N·m at 6800 rpm. A similar approach was recently deployed by Honda with their CBF1000 model.

Compared to the ER-6, the Versys has a redesigned sub-frame, new exhaust header design, and redesigned suspension. The Versys replaces the standard non-adjustable suspension of the ER-6 with fully adjustable upside-down front forks, and at the rear an all alloy swingarm instead of the tubular steel item on the original bike.

KAWASAKI VERSYS

PRODUCTION	2006 - present
MAX SPEED	-
ENGINE TYPE	liquid-cooled V twin
BORE/STROKE	83 x 60 mm
CAPACITY	649cc
COMPRESSION RATIO	10.6:1
TRANSMISSION	6-speed
FINAL DRIVE	Sealed Chain
VALVE TRAIN	DOHC, 8 valve
CARBURETION	Fuel Injection
FRONT TIRE	120/70ZR-17
REAR TIRE	160/60ZR-17
DRY WEIGHT	181 kg
WHEELBASE	1,415 mm
SEAT HEIGHT	840 mm

MOTO GUZZI

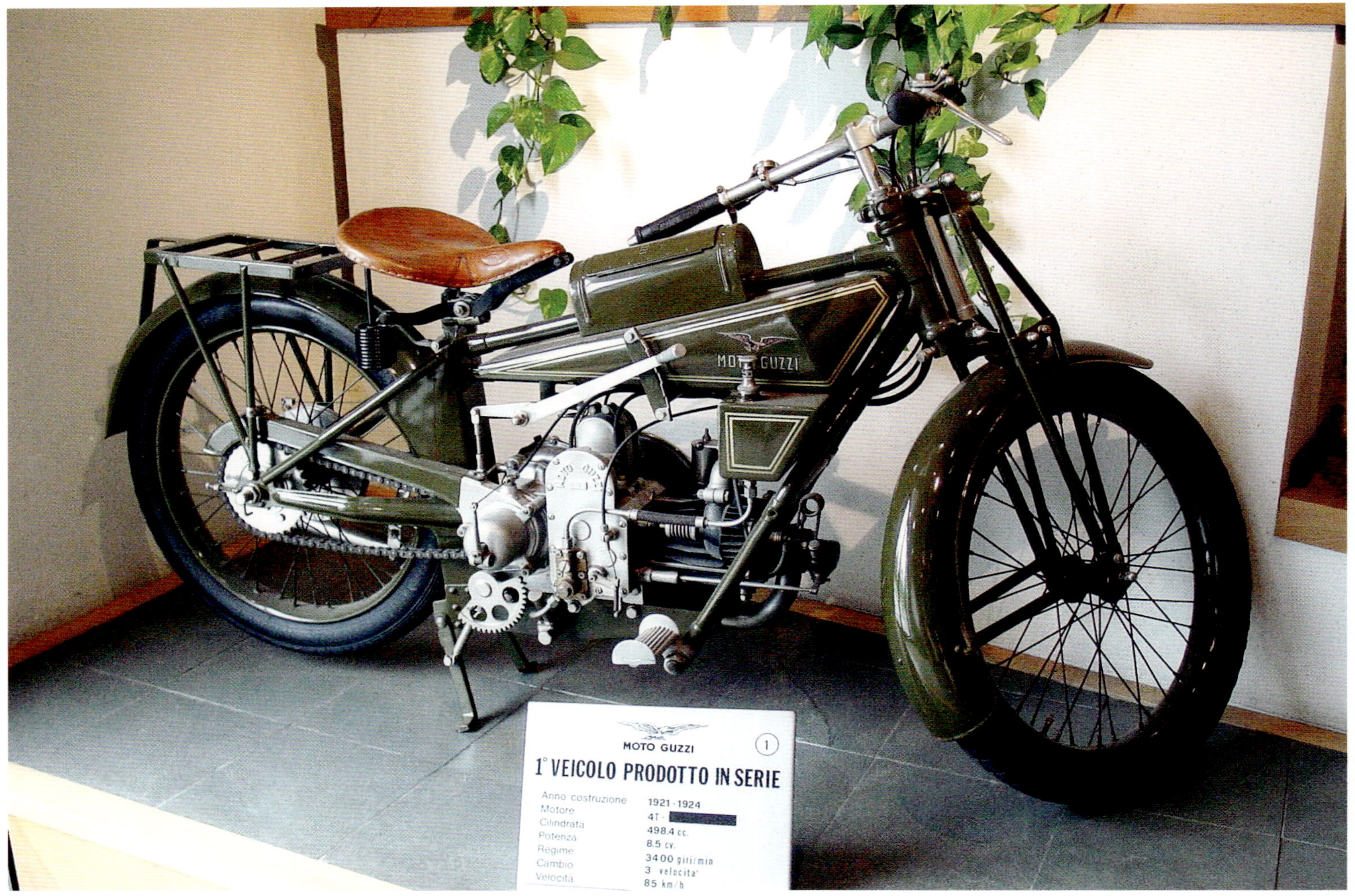

1921 Moto Guzzi Normale

Not unlike Harley-Davidson and other storied motorcycle manufacturers that have survived for decades, Moto Guzzi has experienced a series of business cycles and a series of ownership arrangements – some complex, some brief, some with enduring legacy.

Three themes connect the chapters in Moto Guzzi's history: the importance of racing, the importance of innovation and the ongoing struggle to leverage the company's heritage – the seminal passion for motorcycles that gave rise to the company in 1921.

1921–1966 ~ the origins of Moto Guzzi

Moto Guzzi was conceived by two aircraft pilots and their mechanic serving in the Corpo Aeronautico Militare (the Italian Air Corp, CAM) during World War I: Carlo Guzzi, Giovanni Ravelli and Giorgi Parodi. By happenstance assigned to the same Miraglia Squadron based outside Venice, the three became close, despite starkly different socio-economic backgrounds. The trio envisioned creating a motorcycle company after the war. Guzzi would engineer the motor bikes, Parodi (scion of wealthy Genovese ship-owners) would finance the venture, and Ravelli (already a famous pilot and motocycle racer) would promote the bikes with his racing prowess. Guzzi and Parodi (along with Parodi's brother) formed Moto Guzzi in 1921. Ravelli, ironically, had died just days after the war's end in an aircraft crash – and is commemorated by the eagle's wings that form the Moto Guzzi logo.

Carlo Guzzi and Giorgio Parodi — along with Giorgio's brother Angelo – created a privately held silent partnership "Società Anonima Moto Guzzi" on 15 March 1921 – for the purpose of (according to the original articles of incorporation) "the manufacture and the sale of motor cycles and any other activity in relation to or connected to metallurgical and mechanical industry".

The formation of the company hinged on an initial loan of two thousand Lira from the Parodis' father, Emanuele Vittorio, which he gave on 3 January 1919, offering the balance of the loan upon his review of the project's progress:

"Dear Giorgio, you can let both your partners know that I will offer you for your first 1,500 or 2,000 Lire. Although with the condition that the sum, under no circumstances, shall be increased. Likewise, I reserve the right to supervise your progress before giving my agreement to this project."

The company was legally based in Genoa, Italy, with its headquarters in Mandello. The very earliest motorcycles bore the name G.P. (Guzzi-Parodi), though the marque quickly changed

1941 Moto Guzzi Airone

to Moto Guzzi. As the only actual shareholders, the Parodi's wanted to shield their shipping fortunes by avoiding confusion of name G.P. with Giorgio Parodi's initials. Carlo Guzzi initially received royalties for each motorcycle produced, holding no ownership in the company that bore his name. In 1946 Moto Guzzi formally incorporated as Moto Guzzi S.p.A. with Giorgio Parodi as chairman.

From the 1930s until the 1960s, Moto Guzzi was the largest marque among Italian motorcycle manufacturers. Carlo Guzzi's first engine design was a horizontal single that dominated the first 45 years of the company's history in various configurations. Through 1934, each engine bore the signature of the mechanic who built it. As originally envisioned, the company used racing to promote the brand. In the 1935 Isle of Man TT, Moto Guzzi factory rider Stanley Woods performed an impressive double victory with wins in the Lightweight TT as well as the Senior TT.

Until the mid 1940s, the traditional horizontal four-stroke single cylinder 500 cc engines outfitted with one overhead and one side valve (also known as: IOE, inlet over exhaust or F-head) were the highest performance engines Moto Guzzi sold to the general public. By contrast, the company supplied the official racing team and private racers with higher performance racing machines with varying overhead cam, multi-valve configurations and cylinder designs.

In the 1950s, Moto Guzzi, along with the Italian factories of Gilera and Mondial, led the world of Grand Prix motorcycle racing. With durable and lightweight 250 cc and 350 cc bikes designed by Giulio Carcano, the firm dominated the middleweight classes. The factory won five consecutive 350 cc world championships between 1953 and 1957. In realizing that low weight alone might not continue to win races for the company, Carcano designed the V8 500 cc GP race bike — whose engine was to become one of the most complex engines of its time. Despite the bike's having led many races and frequently posted the fastest lap time, it often failed to complete races because of mechanical problems. Ultimately, the V8 was not developed further as Moto Guzzi withdrew (together with the main competitors Gilera and Mondial) from racing after the 1957 season — citing escalating costs and diminishing motorcycle sales. By the time of its pull out from Grand Prix racing, Moto Guzzi had won 3,329 official races, 8 World Championships, 6 Constructor's Championships and 11 Isle of Man TT victories.

The period after World War II was as difficult in Mandello del Lario as it was elsewhere in post-war Europe. The solution was

1975 Moto Guzzi 750 Ambassador

production of inexpensive, lighter cycles. The 1946 "Motoleggera", a 65 cc lightweight motorcycle became very popular in post-war Italy. A four-stroke 175 cc scooter known as the "Galletto" also sold well. Though modest cycles for the company, the lighter cycles continue to feature Guzzi's innovation and commitment to quality. The step-through Galletto initially featured a manual, foot-shifted three-speed (160 cc) configuration then later a four-speed (175 cc) set-up by the end of 1952. The displacement was increased to 192 cc in 1954 — electric start was added in 1961.

Moto Guzzi was limited in its endeavors to penetrate the important scooter market as motorcycle popularity waned after WWII. Italian scooter competitors would not tolerate an incursion from Moto Guzzi. By innovating the first large-wheeled scooter, Guzzi competed less directly with manufacturers of small-wheeled scooters such as Piaggio (Vespa) and Lambretta. To illustrate the delicate balance within the Italian post-war motorcycle and scooter markets, when Guzzi developed their own prototype for a small-wheeled scooter, Lambretta retaliated with a prototype for a small V-twin motorcycle – threatening to directly compete on Moto Guzzi's turf. The two companies compromised: Guzzi never produced their small-wheeled scooter and Lambretta never manufactured the motorcycle. Notably, the drive train that Lambretta made in their 1953 motorcycle prototype remarkably resembles the V-twin + drive shaft arrangement that Guzzi developed more than ten years later, ultimately to become iconic of the company. The Lambretta Museum in Rodano Italy has one of the two Lambretta prototype motorcycles and the single prototype Guzzi small-wheeled scooter on display.

By 1964, the company was in full financial crisis. Emanuele Parodi and his son Giorgio had died, Carlo Guzzi had retired to private life, and direction passed to Enrico Parodi, Giorgio's brother. Carlo Guzzi died on November 3, 1964 in Mandello, after a brief hospital stay in Davos.

1967–1973 ~ the SEIMM years

In February 1967, SEIMM (Società Esercizio Industrie Moto Meccaniche), a state controlled receiver, took ownership of Moto Guzzi. The SEIMM oversight saw Moto Guzzi adapting to a cultural shift away from motorcycles to automobiles. The company focused on popular lightweight mopeds including the Dingo and Trotter — and the 125 cc Stornello motorcycle. Also during the SEIMM years that Guzzi developed the 90° V twin engine — designed by Giulio Cesare Carcano — which would become iconic of Moto Guzzi.

Though Moto Guzzi has employed engines of myriad configurations, none has come to symbolize the company more than the air-cooled 90° V-twin with a longitudinal crankshaft orientation and the engine's transverse cylinder heads projecting prominently on either side of the bike. The original V-twin was designed in the early 1960s by engineer Giulio Cesare Carcano, designer of the DOHC V8 Grand Prix racer. The air-cooled, longitudinal crankshaft, transverse cylinder, pushrod V-twin began life with 700 cc displacement and 45 hp (34 kW) – designed to win a competition sponsored by the Italian government for a new police bike. The sturdy shaft-drive, air-cooled V-twin won, giving Moto Guzzi renewed competitiveness. This 1967 Moto Guzzi V7 with the original Carcano engine has been continuously developed into the 1200 cc, 80 hp (60 kW) versions offered today (2006). Lino Tonti redesigned the motor for the 1971 Moto Guzzi V7 Sport. This engine is the basis of the currently used 850 cc, 1100 cc and 1200 cc Guzzi engines. Notably, the longitudinal crankshaft and orientation of the engine creates a slight gyroscope effect, with a slightly asymmetrical behavior in turns.

1973–2000 ~ the De Tomaso years

After experiencing financial difficulties in the late 1960s, De Tomaso Industries Inc. (D.T.I. Group or DTI), manufacturer of the De Tomaso sports and luxury cars, owned by Argentinian industrialist Alejandro de Tomaso, purchased SEIMM (and thereby Moto Guzzi) along with Benelli and Maserati in 1973. Under Tomaso's stewardship, Moto Guzzi returned to profitability, though other reports suggest a period of limited investment in Moto Guzzi followed attributed to DTI using Moto Guzzi financially prioritizing their automotive ventures.

In 1979 a small block version of the air-cooled V-twin designed by engineer Lino Tonti was introduced as the V35. Radical when introduced, the design cut the weight from 548 lb (249 kg) of the contemporary 850 T3 to the 385 lb (175 kg) of the V35. The power of the original V35 at 35 bhp (26 kW) was competitive with engines of comparable displacement of the period — later larger versions (V50, V65, V75) were rapidly outclassed by competing water cooled engines. Notably, the Breva and Nevada today feature a descendent of Tonti's V35 engine: the 750 cc V-twin, rated at 48 bhp (36 kW). With its ease of maintenance, durability and even, flat torque curve, the engine design remains suitable to everyday, real-world situations.

As Guzzi continued to develop the V-twin, power was increased in the mid 1980s when Guzzi created 4 valve versions of the "small block" series. Of these, the 650 and the 750 were rated at 60 bhp (45 kW) and 65 bhp (48 kW) respectively. The production of the 4-valve "small block" engines ended in the later 1980s.

Moto Guzzis have used an hydraulic integral brake system, where the right front disc works off the handlebar lever, while the left front and the rear disc work off the foot brake. Rudge-Whitworth used an early integrated, anti-lock, braking system in 1925.

The cartridge front fork used in Guzzi's motorcycles of the later 1970s and 1980s is a Guzzi invention. Instead of containing the damping oil in the fork it is in a cartridge. Oil in the fork is purely for lubrication.

Still under the De Tomaso umbrella, in 1988, Benelli and SEIMM merged to create Guzzi Benelli Moto (G.B.M. S.p.A.). During this period, Moto Guzzi existed as an entity within the De Tomaso owned G.B.M., but in 1996 celebrated its 75th birthday and the return of its name to Moto Guzzi S.p.A. In 1996, De Tomaso became Trident Rowan Group aka TRG.

2000–2004 ~ the Aprilia years

Under the helm of Ivano Beggio, Aprilia S.p.A acquired Moto Guzzi S.p. A on 14 April 2000 for $65 million. According to the original press release, the intention had been that Moto Guzzi would remain headquartered in Mandello del Lario and would share Aprilia's technological, R&D capabilities and financial resources as well. The arrangement would remain short-lived, as Aprilia itself stumbled financially. At the same time Aprilia attempted to diversify in other areas of manufacturing, new Italian laws required helmets for motorcyclists and raising insurance rates for teenage motorcyclists, severely affected the company's profitability. Nonetheless, Aprilia had committed large sums to renovating the Mandello Moto Guzzi factory – renovations that were ultimately completed. Notably, Ducati Motor Holding again made an offer for Moto Guzzi during Aprilia's financial difficulties, as it had before, when Aprilia had purchased Moto Guzzi in 2000. Other potential buyers included Kymco of Taiwan and Bombardier of Canada (Kymco reportedly making the highest offer). The Moto Guzzi assembly line closed for a short period in March of 2004, due to the financial difficulties.

2004 onwards ~ the Piaggio years

On 30 December 2004, Piaggio & Co. S.p.A acquired Aprilia and thereby Moto Guzzi, forming Europe's largest motorycle manufacturer. Moto Guzzi S.p. A officially becomes a Unico Azionista of Piaggio, part of Immsi S.p.A. Investments have allowed introduction of a series of competitive new models in rapid succession. US Moto Guzzi (and Aprilia) dealers experienced considerable parts supply difficulties during the ownership transition, though by summer of 2007, parts supplies were running smoothly.

MOTO GUZZI STELVIO 12004V

Stelvio has been developed with the new "Quattrovalvole", the most advanced evolutionary milestone of the mythical 90° transversal V-twin engine. An engine that has evolved beyond recognition even for the most loyal passionate, and that marvel due to its contents and performance, the highest of the category.

A design that has absolutely revolutionised Mandello's twin-cylinder mechanics, with more than 75% of new components. In all, 563 new parts have been designed to allow this 90° V-twin engine to have a single overhead camshaft distribution system, which controls 4 valves per cylinder.

A great example of mechanics, unique in the motorcycle panorama thanks to its lucid design structure, obtained by light and smooth sliding masses, moving alternately, the thermodynamic study and fuel supply in order to enhance functions, reliability and smooth operation aiming at a new quality standard.

The new lighter crankshaft, 3-ring forged pistons and very light valves with 5-mm stems supported by cone-section springs that eliminate resonance and power losses at high speeds, emerge among the most valuable components present.

Lubrication and cooling are ensured by two pumps fitted in tandem in order to obtain smaller dimensions and are activated by three gears that make lubricant flow along a channel, separated from the cylinder head, to reach the exhaust pipe, where the highest temperature is found, and feed cooling jets placed under the piston. In order to house these new components, a completely new crankcase is necessary, to offer smaller dimensions and an integral front bench support with bushing, while the rear one has a new flange and a more effective oil feeding system. The most seducing element of the new "Quattrovalvole" is, surely, the modern design of cylinders, more compact with cooling finning oriented to riding direction. Consequently, head covers have been redesigned by adding, apart from the "Quattrovalvole" identifying abbreviation, a floating system that further cushions the silent action of the distribution system, controlled by "Morse" chains and supplied with hydraulic tensioners and tensioning pads. Fuel supply has also been subject to important innovations, with throttle bodies well over Ø 50-mm and new IWP 189 injectors.

MOTO GUZZI STELVIO 12004V

PRODUCTION	2006 - present
MAX SPEED	-
ENGINE TYPE	V twin
BORE/STROKE	95 x 81.2 mm
CAPACITY	1,151cc
COMPRESSION RATIO	11: 1
TRANSMISSION	6-speed
FINAL DRIVE	Compact reactive cardan shaft
VALVE TRAIN	-
CARBURETION	Fuel Injection
FRONT TIRE	110/80ZR-19
REAR TIRE	180/55ZR-17
DRY WEIGHT	214 kg
WHEELBASE	2,250 mm
SEAT HEIGHT	840 mm

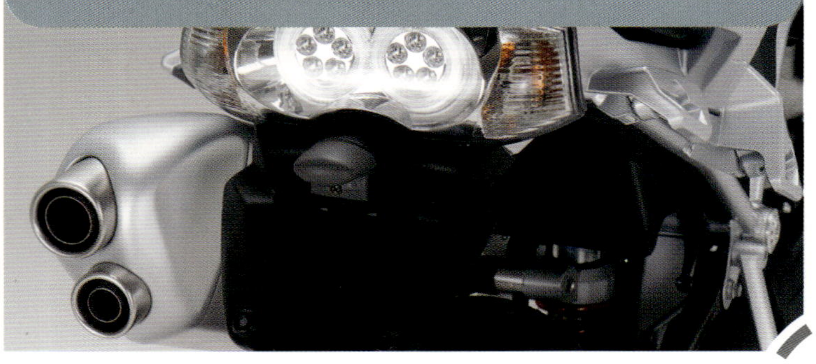

MOTO GUZZI BREVA 1200

The new Breva incorporates all the versatility and easy riding character of Moto Guzzi's touring models to date, with the addition of refined looks, the latest digital technology, and a unique reactive shaft drive system.

The graphics are the same as on competition models, the stylish front discs and the "carbon fibre effect" exhaust that suggest speed give an immediate impression of the 1200 Sport's character.

The most significant technical innovation in the Breva V1200 is reactive shaft drive, patented by the Moto Guzzi under the name 'CA.R.C.' ('Cardano Reattivo Compatto' - Compact Reactive Shaft Drive).

The main benefit of this original and extremely compact drive system is the elimination of the counter-shaft effect typical of conventional shaft drives. Unlike alternative systems, Moto Guzzi's patented reactive shaft drive enables a single piece swing arm to be used. Because the reaction shaft is not load-bearing, it is perfectly reliable in operation. The new system ensures a consistently smooth ride, without the jerks on acceleration or deceleration that are inevitable with conventional shaft drives.

The new system maintains all the other advantages of shaft drive including cleanliness, low noise levels and very low maintenance.

MOTO GUZZI BREVA 1200	
PRODUCTION	2006 - present
MAX SPEED	-
ENGINE TYPE	V twin
BORE/STROKE	95 x 81.2 mm
CAPACITY	1,151cc
COMPRESSION RATIO	9.8 : 1
TRANSMISSION	6-speed
FINAL DRIVE	Compact reactive cardan shaft
VALVE TRAIN	-
CARBURETION	Fuel Injection
FRONT TIRE	120/70ZR-17
REAR TIRE	180/55ZR-17
DRY WEIGHT	229 kg
WHEELBASE	2,195 mm
SEAT HEIGHT	800 mm

MOTO GUZZI BELLAGIO

The appearance of the Bellagio is that of a pure custom machine with a low two-up rear-set seat sitting astride the rear wheel, forward positioned footpegs and spectacular pulled-back handlebars for complete control. Surrounded by the music from the exhaust, enjoying the comfort provided by the saddle that was specially designed for long journeys and revelling in the superb suspension, the lone rider soon encounters the essence of motorcycling aboard this bike.

The evenly distributed weight provides surprising manoeuvrability and even challenging conditions such as city traffic hold no fears, thanks to the ease of changing direction aboard the bike, and its weighted front end. Then, the sheer pleasure of riding will see you lose yourself amidst all the little details of its seductive style, when you can truly appreciate the finishing touches and enjoy the softness of the controls. It is at moments like this when the rider is in complete harmony with the machine, when he feels in touch with the spirit of an engine that can take him up to enormous speed or carry him along at a snails pace past the mirrored walls of the city. The elegant instrument panel is another key element to the Bellagio. It gracefully combines an analogue speedometer with a white face and numerals with a high-tech LCD on-board trip computer that allows the rider to be in complete control of the bike and his destiny through comfortably positioned commands.

MOTO GUZZI BELLAGIO	
PRODUCTION	2006 - present
MAX SPEED	-
ENGINE TYPE	V twin
BORE/STROKE	95 x 66 mm
CAPACITY	935.6cc
COMPRESSION RATIO	10 : 1
TRANSMISSION	6-speed
FINAL DRIVE	Compact reactive cardan shaft
VALVE TRAIN	-
CARBURETION	Fuel Injection
FRONT TIRE	120/70ZR-17
REAR TIRE	180/55ZR-17
DRY WEIGHT	224 kg
WHEELBASE	2,270 mm
SEAT HEIGHT	780 mm

MOTO GUZZI GRISO 8V

Don't let the long low profile fool you: The Griso 8V has handling that would not shame a supersport machine. The secret lies in the high-tensile steel frame with twin upper supports whose design owes little to aesthetics and much to sophisticated engineering concepts that provide exceptional rigidity under moments of extreme torsional stress. The tubular steel twin cradle frame is connected to a single-sided aluminium rear swingarm housing the CARC system.

The frame geometry had been carefully calculated to provide stability with the steering angle set at 26° and rake at 108 mm. The wheelbase measures 1544 mm and the Griso 8V is surprisingly manoeuvrable, even lightning fast in changes of direction.

These handling characteristics are due to the low centre of gravity and a rock solid off-set steering head that provides an immediate response to pressure on the handlebars. The front end is glued to the road thanks to upside down forks from racing stock. The 43 mm forks are adjustable for rebound and compression. A Boge progressive suspension unit is fitted to the rear and this too had three classic settings. The sophisticated suspension ensures a neutral set-up on entering and exiting corners, under acceleration from the 110 CV "Quattrovalvole" engine and under braking from the Brembo brake system. These brakes are both aesthetically pleasing and highly effective with their radial calipers acting on "wave" discs that have been specially designed for maximum heat dispersion.

MOTO GUZZI GRISO 8V	
PRODUCTION	2007 - present
MAX SPEED	-
ENGINE TYPE	V twin
BORE/STROKE	95 x 81.2 mm
CAPACITY	1,151cc
COMPRESSION RATIO	11 : 1
TRANSMISSION	6-speed
FINAL DRIVE	Compact reactive cardan shaft
VALVE TRAIN	-
CARBURETION	Fuel Injection
FRONT TIRE	120/70ZR-17
REAR TIRE	180/55ZR-17
DRY WEIGHT	222 kg
WHEELBASE	2,260 mm
SEAT HEIGHT	800 mm

MOTO GUZZI NORGE 1200

Five thousand kilometres in six days with not a shadow of a problem mark the Moto Guzzi Norge 1200 GT's as the élite of "gran turismo" motorcycles. The Norges are superbly reliable machines and instantly found willing partners in the fourteen international testers that rode them as far as North Cape following the same tracks as the G.T.

500 Norge ridden by Giuseppe "Naco" Guzzi on his way to the Artic Circle in 1928. This epic journey is to celebrate 85 years of Moto Guzzi and the festivities will conclude on the 15th. 16th and 17th September with the GMG-"Giornate Mondiali Guzzi" (World Guzzi Days) at Mandello del Lario.

The latest addition to the Moto Guzzi range is the Norge 1200, a motorcycle that embodies all of Moto Guzzi's traditional long distance touring values, right down to the name itself. The Norge 1200, which was first presented to the world at Eicma 2005, is now available in four different versions. The machines of Mandello del Lario are therefore once again the rightful protagonists of Gran Turismo, a genre that owes its very existence to the genius of the Guzzi brothers. For countless years in the past, Moto Guzzi set the standards for touring comfort and technology. Now, after a gap of twenty years, Moto Guzzi is back, and determined to regain its position of leadership on the tourer market.

MOTO GUZZI NORGE 1200	
PRODUCTION	2005 - present
MAX SPEED	-
ENGINE TYPE	V twin
BORE/STROKE	95 x 81.2 mm
CAPACITY	1,151cc
COMPRESSION RATIO	9.8 : 1
TRANSMISSION	6-speed
FINAL DRIVE	Compact reactive cardan shaft
VALVE TRAIN	-
CARBURETION	Fuel Injection
FRONT TIRE	120/70ZR-17
REAR TIRE	180/55ZR-17
DRY WEIGHT	246 kg
WHEELBASE	1,495 mm
SEAT HEIGHT	800 mm

PAIGGIO

1946 Vespa 98cc

Piaggio based in Pontedera, Italy encompasses seven brands producing scooters and motorcycles. As the fourth largest producer of scooters and motorcycles in the world, Piaggio produces more than 600,000 vehicles annually, with five Research and development centers, more than 6,700 employees and operations in over 50 countries.

Founded by Rinaldo Piaggio in 1884, Piaggio initially produced locomotives and railway carriages. During World War I the company focused on producing airplanes.

During World War II the company produced fighter planes, but Piaggio emerged from the conflict with its Pontedera plant completely demolished by bombing. Italy's crippled economy and the disastrous state of the roads did not assist in the re-development of the automobile markets. Enrico Piaggio, the son of Piaggio's founder Rinaldo Piaggio, decided to leave the aeronautical field in order to address Italy's urgent need for a modern and affordable mode of transportation. The idea was to design an inexpensive vehicle for the masses.

Aeronautical engineer Corradino D'Ascanio, responsible for the design and construction of the first modern helicopter by Agusta, was given the job of designing a simple, robust and affordable vehicle by Enrico Piaggio. The vehicle had to be easy to drive for both men and women, be able to carry a passenger, and not get its driver's clothes dirty. Consequently, in 1946 Piaggio launched the Vespa (Italian for "wasp") scooter, and within 10 years over a million units had been produced.

With strong cash flow emanating from the success of the Vespa, Piaggio developed other products, including the 1957 Vespa 400, a tiny passenger car.

In 1959, Piaggio came under the control of the Agnelli family, the owners of car maker Fiat SpA. Resultantly, as the wider ownership of Fiat in Italian industry, in the 1964 the two divisions (aeronautical and motorcycle) split to become two independent companies; the aeronautical division was named IAM Rinaldo Piaggio. Today the airplane-company Piaggio Aero is controlled by the family of Piero Ferrari, who also still holds 10% of the famous car maker Ferrari.

Vespa thrived, until 1992 when Giovanni Alberto Agnelli became CEO - but Agnelli was already suffering from cancer, and died in 1997. In 1999, Morgan Grenfell Private Equity acquired Piaggio, but a quickly hoped for sale was dashed by a failed joint venture in China. In Italy, Piaggio invested 15 million euros ($19.4 million) in a new motorcycle but dropped it after building a prototype. By the end of 2002, the company had run up 577 million euros in debt on revenues of 945 million euros, and booked a loss of 129 million euros.

Then came Roberto Colaninno, "A lot of people told me I was crazy. Piaggio wasn't dying. It just needed to be treated better." said Colaninno. Piaggio's finances where in a bad shape, but

1976 Vespa 125 Primavera ET3

its brand was still well known and its products were featuring in more Hollywood films thanks to the Vespa ET4. In 1995, Colaninno had pulled off what was then Europe's largest-ever hostile takeover when he took control of Telecom Italia SpA. In October 2003, Mr. Colaninno made an initial investment of 100 million euros through his holding company Immsi SpA in exchange for just under a third of Piaggio and the mandate to run it. Chief executive Rocco Sabelli, redesigned the factory to Japanese principles, and redesigned the factory so that every Piaggio scooter could be made on any assembly line.

Unlike the turnaround recipe applied at U.S. auto makers, Mr. Colaninno didn't fire a single worker - a move which helped seduce the company's skeptical unions. "Everyone in a company is part of the value chain," said Colaninno. All bonuses for blue-collar workers and management were based on the same criteria: profit margins and customer satisfaction; and air conditioning was installed in the factory. He also gave the company's engineers, who had been idled by the company's financial crisis, deadlines for projects - they rolled out two world firsts in 2004: a gas-electric hybrid scooter; a scooter with two wheels in front and one in back which grips the road better.

One of Piaggio's problems Mr. Colaninno couldn't fix from the inside was its scale. Even though Piaggio was the European market leader, it was dwarfed by rivals Honda and Yamaha. A year after restoring Piaggio's health, Colaninno directed Piaggio's takeover of the Italian scooter and motorcycle manufacturer Aprilia, and with it the Aprilia-owned Moto Guzzi, storied Italian manufacturer of motorcycles.

On 23 April, 1946 at 12 o'clock in the central office for inventions, models and makes of the Ministry of Industry and Commerce in Florence, Piaggio e C. S.p.A. took out a patent for a "motorcycle of a rational complexity of organs and elements combined with a frame with mudguards and a casing covering the whole mechanical part".

The basic patented design allowed a series of features to be deployed on the spar-frame, which would later allow quick development of new models. The original Vespa featured a rear pillion seat for a passenger, or optionally a storage compartment. The original front protection "shield" was a flat piece of aero metal; later this developed in to a twin skin to allow additional storage behind the front shield, similar to the glove compartment in a car. The fuel cap was located underneath the (hinged) seat, which saved the cost of an additional lock on the fuel cap or need for additional metal work on the smooth skin.

PIAGGIO BEVERLY TOURER

The Beverly Tourer is the new entry in the Beverly range for 2008, with 400, 250 and 125 cc engines. A technologically advanced performer, the Beverly Tourer is characterized by the sleek lines of a high wheel scooter, sure to give the rider all the thrills of a motorcycle ride, while remaining in full control of the vehicle.

Perfect for every-day use, the Beverly Tourer meets the widest range of requirements for safe mobility, completely at its ease in the metropolitan environment where its dynamic qualities and distinctive elegance are notable to say the least.

The refined finish and the meticulous attention to detail make the Beverly Tourer really stand out from the crowd: the frontal has been completely redesigned, in line with its "big brother" the Cruiser. The headlight is mounted on the handlebars with an elegant windscreen made of tinted methacrylate above the bars, to offer the rider more protection.

The leg shield has a new radiator grill with a chromed finish, giving the Beverly Tourer the looks of a real classic.

The elegant Piaggio "family feeling" style has been embellished with chromed finishes that characterise the lateral trim on the leg shield and on the rear side panels. The rear-view mirrors are also chromed and match the new headlight-handlebars design, giving the Beverly Tourer a sophisticated and important image. The exhaust heatshield on the 250 and 125 cc version is chromed. Other changes include the new front mudguard and the rear light, which is now clear. The finish of the seat is also new with contrasting stitching, while the storage compartment under the seat can hold two half-jet helmets.

The instruments have been redesigned with new graphics in line with the new classic Beverly Tourer image, reasserting a stylistic coherence, which has been the underlying theme in planning the new Beverly family. Also the choice of the paint has been studied to exalt the elegance of the lines and finishes on the Beverly Tourer. The Beverly Tourer comes in a choice of four colours: Pulsar Grey, Midnight Blue, Naiade Ivory and Graphite Black.

PIAGGIO BEVERLY TOURER

PRODUCTION	2001 - present
MAX SPEED	155 km/h
ENGINE TYPE	Master Single Cylinder
BORE/STROKE	85.8 x 69 mm
CAPACITY	399cc
COMPRESSION RATIO	10.5:1
TRANSMISSION	Continuous Variable Transmission
FINAL DRIVE	-
VALVE TRAIN	SOHC; 4 valve
CARBURETION	Fuel Injection
FRONT TIRE	110/70ZR-16
REAR TIRE	150/70ZR-14
DRY WEIGHT	198 kg
WHEELBASE	1,550 mm
SEAT HEIGHT	775 mm

PIAGGIO CARNABY

Combining good performance and compact size, the Piaggio Carnaby makes a great first impression with its pleasing looks and the instantly comfortable, welcoming feeling it gives you when you get on. It has a gentle, expressive and modern shape. You'll immediately notice how comfortable the flat footrest platform is for city use. The protective front shield holds an innovative and stylish headlamp unit that matches the Carnaby's modern lines to perfection.

The 2008 Carnaby features a generous new seat upholstered in a new non-slip material for even greater comfort. At a height of only 790 mm, the seat is a key feature of the Carnaby's perfect ergonomics, assisted by the scooter's broad handlebars. A new tinted windshield is mounted on the handlebars to give the 2008 Carnaby an even more elegant look while improving aerodynamic penetration too.

Even the instruments have been updated. The surround is now paint finished to match the scooter's bodywork in a style evocative of the world of prestige automobiles. A new chrome plated "Carnaby" logo on the handlebars adds to the elegance of the instrument panel.

The Carnaby has also renewed its colour schemes for 2008, and is now available in four elegant finishes: metallic Graphite Black, Pulsar Grey and Andromeda Blue and pastel shade Naiade Ivory.

Everything about the Carnaby conveys a sensation of practicality, strong personality and a no-nonsense, functional image. In short, the Carnaby is the perfect solution to city traffic. The Carnaby is built around two technically advanced Piaggio Leader engines, a 200cc and a 125 cc unit. These latest-generation engines are fully compatible with the latest and strictest Euro 3 emission standards. Both engines are identical in configuration: 4 stroke, 4 valve, and liquid cooled, and both offer pace-setting power and performance. The 125 cc develops 15 HP, while the 200 cc delivers 21 HP. Generous power is always on tap, right through the speed range, to guarantee a consistently smooth and enjoyable ride and extremely low consumption levels.

PIAGGIO CARNABY

PRODUCTION	2007 - present
MAX SPEED	104 km/h
ENGINE TYPE	Single Cylinder
BORE/STROKE	57 x 48.6 mm
CAPACITY	124cc
COMPRESSION RATIO	12.5:1
TRANSMISSION	Continuous Variable Transmission
FINAL DRIVE	-
VALVE TRAIN	-
CARBURETION	Fuel Injection
FRONT TIRE	110/70ZR-16
REAR TIRE	130/70ZR-16
DRY WEIGHT	146 kg
WHEELBASE	1,365 mm
SEAT HEIGHT	790 mm

PIAGGIO LIBERTY 125

Contemporary and sleek, the Liberty 125 is equipped with a Piaggio Leader 125cc 4-stroke engine offering smooth acceleration and excellent economy. Features such as the large wheels, standard electronic immobiliser, wide footrest, underseat storage, low and comfortable seat and clear instrument panel guarantee convenience, high levels of grip and everyday practicality. Light and easy to ride, the Liberty 125 is perfect for inner-city, suburban and longer-range journeys.

PIAGGIO LIBERTY 125

PRODUCTION	2001 - present
MAX SPEED	-
ENGINE TYPE	Single Cylinder
BORE/STROKE	57 x 48.6 mm
CAPACITY	124cc
COMPRESSION RATIO	12.5:1
TRANSMISSION	Continuous Variable Transmission
FINAL DRIVE	-
VALVE TRAIN	SOHC; 2 valve
CARBURETION	Fuel Injection
FRONT TIRE	90/80ZR-16
REAR TIRE	110/80ZR-16
DRY WEIGHT	-
WHEELBASE	1,325 mm
SEAT HEIGHT	805 mm

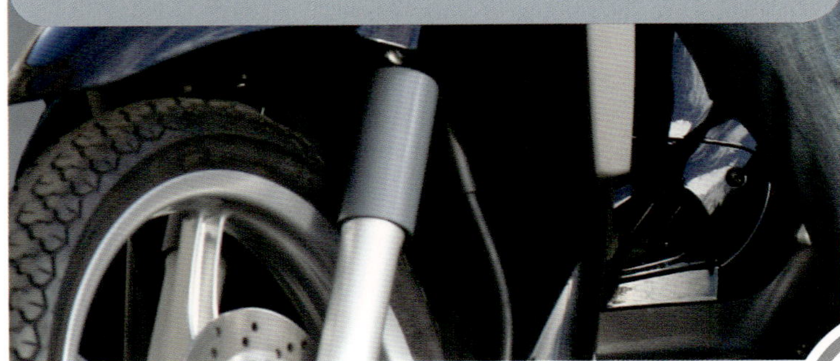

PIAGGIO VESPA S

Sport, dynamism and originality are the trademarks of the all-new Vespa S. With its clean, essential lines, the Vespa S is a minimalist version of the legendary seventies models, and proof that the young spirit of Vespa is still very much alive.

The handlebar fairing houses a new rectangular headlight. This, of course, is not the first time for a Vespa to feature a rectangular headlight: fans are sure to recall the amazing 50 Special, a favourite with teenagers throughout the fabulous seventies. The front shield is more impressive too, with a new, aggressive looking air intake.

Reduced in size to reveal the layout of the suspension, the new shape mudguard also features a stylish chrome trim for an even sleeker look. The small size of the mudguard leaves the wheel and light alloy rim in full view in a tribute to the performance and high-tech design of the Vespa S.

The element that has seen most change and that characterises the new design is the front shield. Its clean styling recovers the pure lines that have always been the Vespa's trademark. The shield's uncluttered surface is fundamental and its simplicity and minimal thickness powerful design features.

The seat is typically '70s in styling and is available in two versions: the 'Sport', fitted as standard to the Vespa S 50 to enhance its dynamic character, and the 'Touring', fitted to the 125 to maximise comfort and usability. Both are available as options and can be fitted to either model. The seat is upholstered in new materials and enhanced with a classy trim that emphasises its stylish shape.

The rear of the Vespa S has an all-new look with a sleek, simple look and a new tail light designed especially to enhance the dynamism of this sporty model.

PIAGGIO VESPA S

PRODUCTION	2007 - present
MAX SPEED	91km/h
ENGINE TYPE	Single Cylinder
BORE/STROKE	57 x 48.6 mm
CAPACITY	124cc
COMPRESSION RATIO	-
TRANSMISSION	Continuous Variable Transmission
FINAL DRIVE	-
VALVE TRAIN	SOHC; 2 valve
CARBURETION	Fuel Injection
FRONT TIRE	110/70ZR-11
REAR TIRE	120/70ZR-10
DRY WEIGHT	110kg
WHEELBASE	1,280 mm
SEAT HEIGHT	785 mm

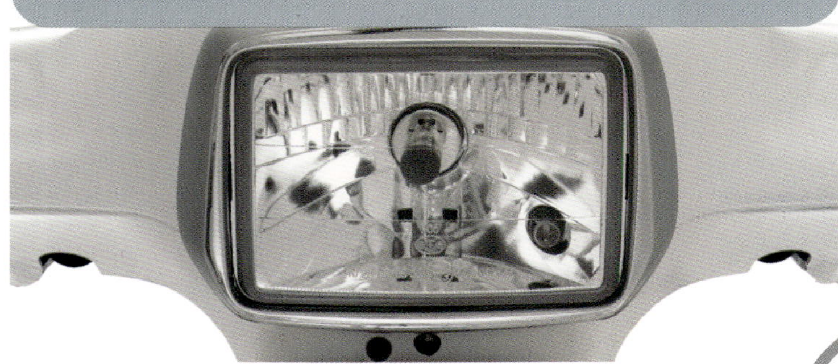

TRIUMPH

TRIUMPH

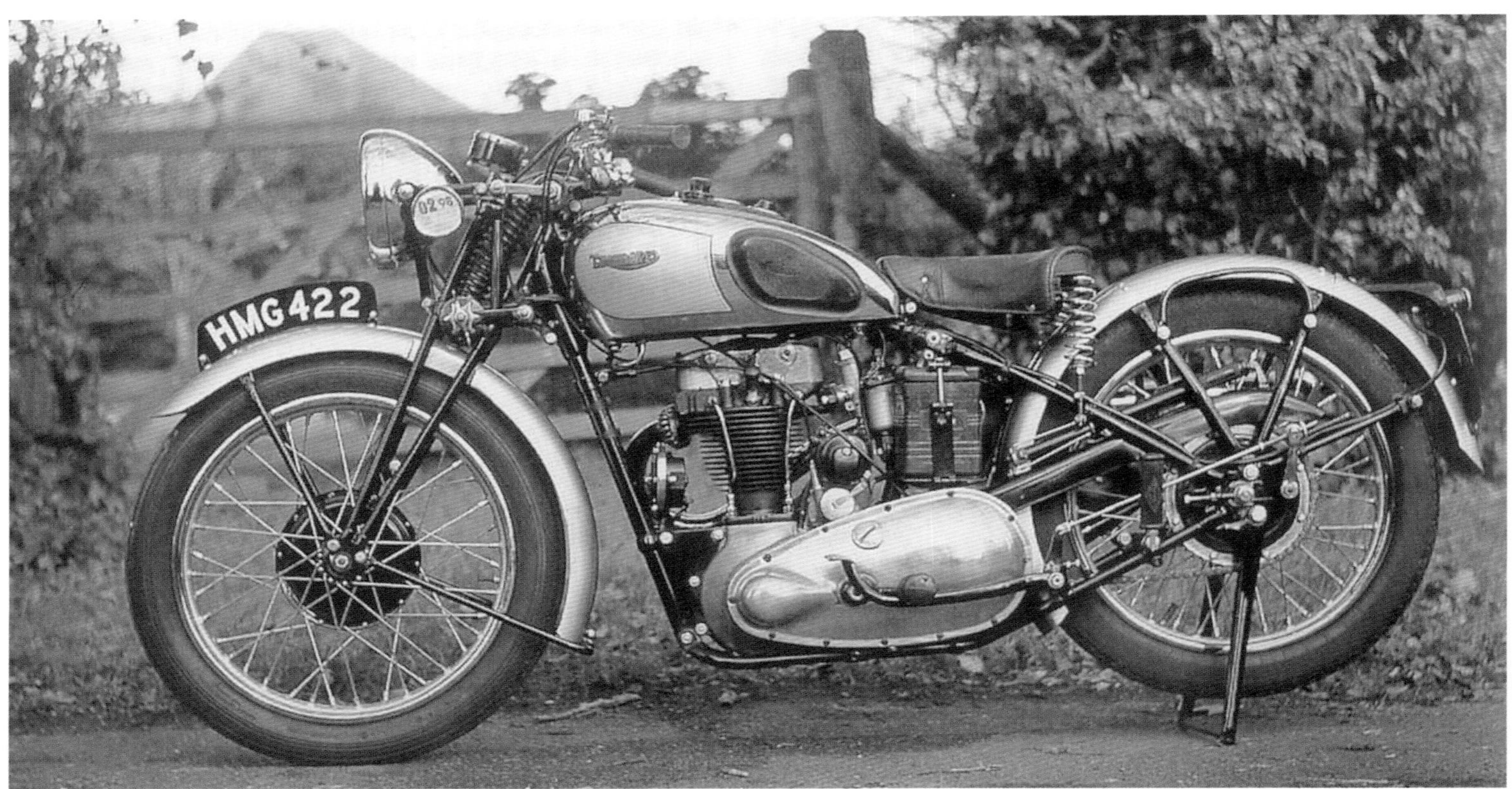

1939 Triumph Tiger

The company began in 1885 when Siegfried Bettmann emigrated to Coventry in England from Nuremberg, part of the German Empire. In 1884 aged 20, Bettmann founded his own company, the S. Bettmann & Co. Import Export Agency, in London. Bettmann's original products were bicycles, which the company bought and then sold under its own brand name. Bettmann also distributed sewing machines imported from Germany.

In 1886, Bettmann sought a more universal name, and the company became known as the Triumph Cycle Company. A year later, the company registered as the New Triumph Co. Ltd., now with financial backing from the Dunlop Pneumatic Tyre Company. In that year, Bettmann was joined by another Nuremberg native, Moritz Schulte.

Schulte encouraged Bettmann to transform Triumph into a manufacturing company, and in 1888 Bettmann purchased a site in Coventry using money lent by his and Schulte's families. The company began producing the first Triumph-branded bicycles in 1889. In 1896, Triumph opened a subsidiary, Orial TWN (Triumph Werke Nuremberg) a German subsidiary for cycle production in his native city.

In 1898, Triumph decided to extend its own production to include motorcycles and by 1902, the company had produced its first motorcycle - a bicycle fitted with a Belgian-built engine. In 1903, as its motorcycle sales topped 500, Triumph opened motorcycle production at its unit in Germany. During its first few years producing motorcycles, the company based its designs on those of other manufacturers. In 1904, Triumph began building motorcycles based on its own designs and in 1905 produced its first completely in-house designed motorcycle. By the end of that year, the company had produced more than 250 of that design.

In 1907, after the company opened a larger plant, production reached 1,000 bikes. Triumph had also launched a second, lower-end brand, Gloria, produced in the company's original plant.

World War I

The outbreak of World War I proved a boost for the company as production was switched to support the Allied war effort. More than 30,000 motorcycles - among them the Model H Roadster aka the "Trusty Triumph," often cited as the first modern motorcycle - were supplied to the Allies.

Bettmann and Schulte fell out after the war, with Schulte wishing to replace bicycle production with cars. Schulte left the company, but in the 1920s Triumph purchased the former Hillman car factory in Coventry and produced a saloon car in 1923 under the name of the Triumph Motor Company.

By the mid-1920s Triumph had grown into one of Britain's leading motorcycle and car makers, with a 500,000 square feet (46,000 m^2) plant capable of producing up to 30,000 motorcycles and cars each year. Triumph also found its bikes in high demand overseas, and export sales became a primary source of the company's revenues, although for the United States, Triumph models were manufactured under license. The company found its first automotive success with the debut of the Super Seven car in 1928. Shortly after, the Super Eight was born.

1930s

When the Great Depression hit in 1929, Triumph spun off

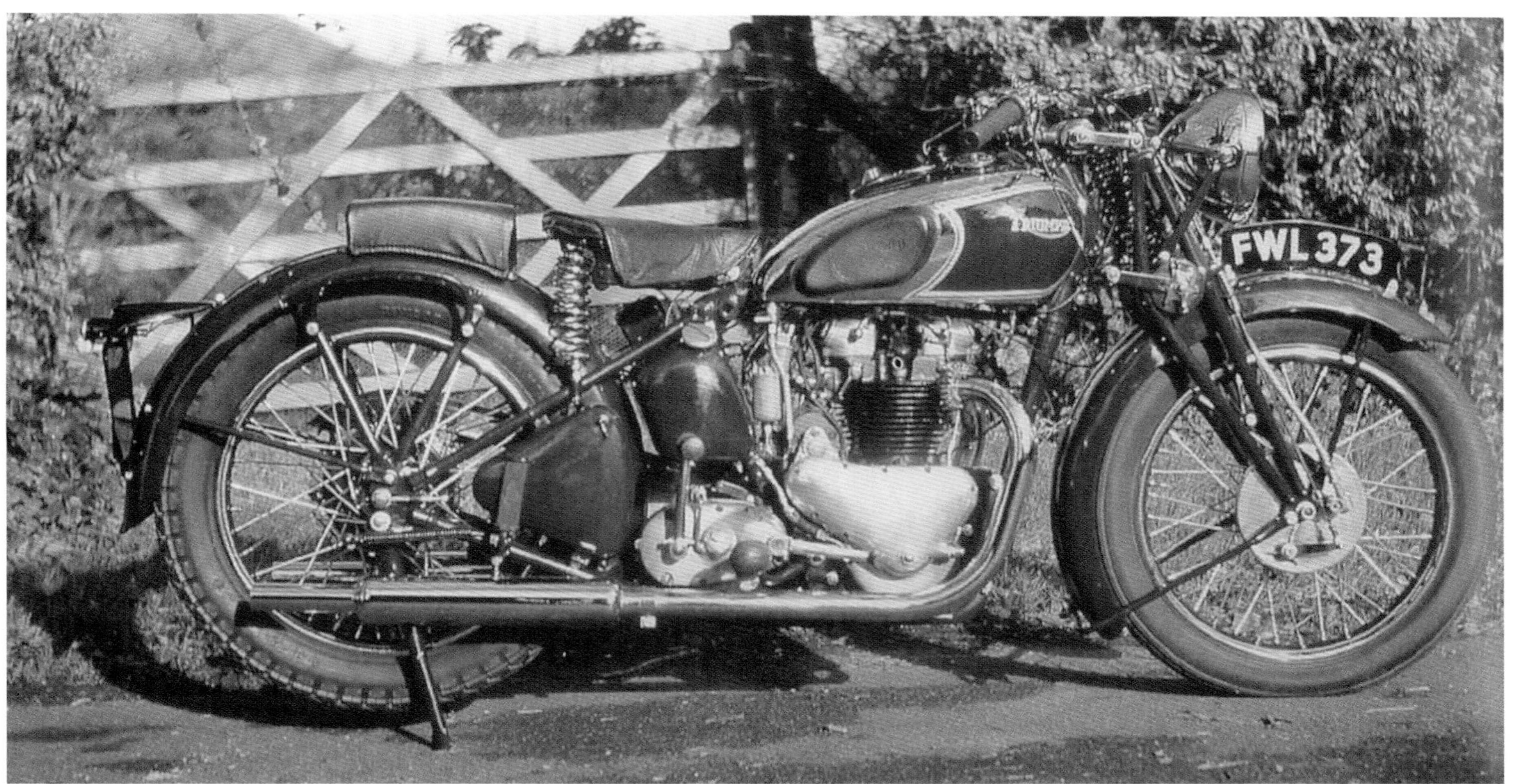

1937 Triumph Speeedtwin

its German subsidiary as a separate, independently owned company, which became part of the Triumph-Adler Company. The Nuremberg firm continued to manufacture motorcycles under the Triumph brand until 1957. In 1932, Triumph sold off another part of the company, its bicycle manufacturing facility to Raleigh. By then, Triumph had been struggling financially, and Bettmann had been forced out of the chairman's spot. He retired completely in 1933

In 1936, the company's two components became separate companies. Triumph always struggled to make a profit from cars, and after going bankrupt in 1939 was acquired by the Standard Motor Company. The motorcycle operations fared better, having been acquired in 1936 by Jack Sangster, who also owned the rival Ariel motorcycle company. That same year, the company began its first exports to the United States, which quickly grew into the company's single most important market. Sangster's formed the Triumph Engineering Co Ltd largely led by ex-Ariel employees, including Edward Turner who designed the 500 cc 5T Speedtwin - released in September 1937, and the basis for all Triumph twins until the 1980s. In 1939 the 500 cc T100 Tiger, capable of 100 miles per hour (160 km/h), was released, and then the war began.

World War II

Motorcycles were produced at Coventry until World War II. The town of Coventry was virtually destroyed in The Blitz (September 7, 1940 to May, 1941). Tooling and machinery was recovered from the site of the devastation and production restarted at the new plant at Meriden, West Midlands in 1942. One of Triumph's wartime products is of particular interest: portable generators for the RAF, using 500 cc Triumph engines with alloy barrels.

Post-war era

The Speed Twin designed by Edward Turner before the war was produced in large numbers after the war. Efforts to settle the lend-lease debts caused nearly 70% of Triumphs post war production to be shipped to the United States.

Post War, the Speed Twin and Tiger 100 were available with a sprung rear hub, Triumph's first attempt at a rear suspension.

Privateers put wartime surplus alloy barrels on their Tiger 100 racers, and won races, inspiring the Triumph GP model. By 1950 the supply of barrels was exhausted, and the GP model was dropped. The American market applied considerable pressure to reverse this backward step, and a die cast close finned alloy barrel was made available. The alloy head made the valve noise more obvious, so ramp type cams were introduced for alloy head models to reduce the noise.

Another motorcycle based on the wartime generator engine was the 499 cc TR5 Trophy Twin, also introduced at the 1948 Motor Cycle Show. It used a single carburettor, low compression version of the Grand Prix engine. Britain won the prestigious 1948 International Six Days Trial. The Triumph works team had finished unpenalised. One team member, Allan Jefferies, had been riding what amounted to a prototype version.

To satisfy the American appetite for motorcycles suited to long distance riding, Turner built a 650 cc version of the Speed Twin design. The new bike was named the Thunderbird (A name

Triumph T140D

Triumph would later license to the Ford Motor Company for use on a car). Only one year after the Thunderbird was introduced a hot rodder in Southern California mated the 650 Thunderbird with a twin carb head originally intended for GP racing and named the new creation the Wonderbird. That 650 cc motor, designed in 1939, held the world's absolute speed record for motorcycles from 1955 until 1970.

The Triumph brand received considerable publicity in the United States when Marlon Brando rode a 1950 Thunderbird 6T in the 1953 motion picture, The Wild One.

The Triumph Motorcycle concern was sold to their rivals BSA by Sangster in 1951. This sale included Sangster becoming a member of the BSA board. Sangster was to rise to the position of Chairman of the BSA Group in 1956.

The production 650 cc Thunderbird was a low compression tourer, and the 500 cc Tiger 100 was the performance bike. That changed in 1954, with the change to swing arm frames, and the release of the alloy head 650 cc Tiger 110, eclipsing the 500 cc Tiger 100 as the performance model.

In 1959, the T120, a tuned double carburettor version of the T110, came to be called the Bonneville. As Triumph and other marques gained market share, Harley became aware that their 1 litre-plus bikes were not as sporty as the modern rider would like, resulting in a shrinking share of the market. The Triumphs were models for a new, "small" Harley Davidson as a result: the now-fabled Sportster, which started out as Harley's version of a Triumph Bonneville. With its anachronistic V-twin, the Sportster was no match for the Bonneville, but it proved a solid competitor in US sales and eventually also in longevity.

In the 1960s, despite internal opposition from those who felt that it would dilute the macho image of the brand, Triumph produced two scooters; the Triumph Tina, a small and low performance 2-stroke scooter of around 100 cc with automatic clutch and a handlebar carry basket, and the Triumph Tigress, a more powerful scooter available with either a 175 cc 2-stroke single or a 250 cc 4-stroke twin engine for the enthusiast.

In 1962, the last year of the "pre-unit" models, Triumph used a frame with twin front down-tubes, but returned to a traditional Triumph single front downtube for the unit construction models that followed. The twin down tube, or duplex frame, was used on the 650 twins, as a result of frame fractures on the Bonneville.

Introduced in 1959, for the 1960 model year, it soon needed strengthening, and was dropped in 1962, with the advent of the unit engines for the 650 range. The 3TA (21) was the first unit construction twin, soon followed by the short-stroke, 490 cc "500" range.

From 1963 all Triumph engines were of unit construction. In 1969 Malcolm Uphill, riding a Bonneville, won the Isle of Man Production TT with a race average of 99.99 miles per hour (160.92 km/h) per lap, and recorded the first ever over 100 miles per hour (161 km/h) lap by a production motorcycle at 100.37 miles per hour (161.53 km/h). For many Triumph fans, the 1969 Bonneville was the best Triumph ever. American sales had already peaked, in 1967. In truth, the demand for motorcycles was rising, but Triumph could not keep up.

In the 1960s, 60% of all Triumph production was exported, which, along with the BSA's 80% exports, made the group susceptible to the Japanese expansion. By 1969 fully 50% of the US market for bikes over 500 cc belonged to Triumph, but technological advances at Triumph had failed to keep pace with the rest of the world. Triumphs lacked electric start mechanisms, relied on push-rods rather than overhead cams, vibrated noticeably, often leaked oil, and had antiquated electrical systems; while Japanese marques such as Honda were building more advanced features into attractive new bikes that sold for less than their British competitors. Triumph motorcycles as a result were nearly obsolete even when they were new; further, Triumph's manufacturing processes were highly labour-intensive and largely inefficient. Also disastrous, in the early 1970s the US government arbitrarily mandated that all motorcycle imports must have their shift and brake pedals in the Japanese configuration, which required expensive retooling of all the bikes for US sale.

The British marques were poorly equipped to compete against the massive financial resources of Japanese heavy industries that targeted competitors for elimination via long-term plans heavily subsidized by the Japanese government. Triumph and BSA were well aware of Honda's ability but while the Japanese were only making smaller engined models, the large engine market was considered safe. When the first Honda 750 cc four cylinder was released for sale to the public, Triumph and BSA were facing trouble. A 3-cylinder engined motorcycle was developed to compete against the Japanese fours: the BSA Rocket 3/Triumph Trident.

The 1970 Tiger/Bonneville re-design and taller twin front downtube oil tank frame met a mixed reception from Triumph enthusiasts at the time, and was insufficient to win back those already riding the Japanese bikes that had hit the markets in 1969; the Honda 750 Four, and the Kawasaki 500 Mach 3. The Triumph 350 cc Bandit received pre-publicity, before being quietly shelved. Triumph was still making motorcycles, but they no longer looked like the bikes Triumph fans expected. The Trident attracted its own market, but the Japanese bikes were improving more rapidly.

Harley Davidson had responded to Triumph's earlier marketing success by producing sportier models that retained the engine design traditional Harley owners identified with, and had managed to survive. Triumph did not manage to do as well with its redesign. Problems were compounded in 1970 by difficulties with parts supply and the labour force. In 1971 a five speed gearbox was introduced.

The parent BSA group made losses of 8.5 million pounds in 1971, 3 million for BSA motorcycles alone. The British government became involved. The company was sold to Manganese Bronze Holdings, which also owned Norton, AJS, Matchless, Francis-Barnett, James-Velocette and Villiers. A new company called Norton Villiers Triumph (NVT), managed by Dennis Poore, emerged.

NVT collapse

When the BSA group collapsed under its debts, government help led to a merger with the Manganese Bronze subsidiary Norton-Villiers. The three remaining brands to be produced by the company were combined to create the new group name of Norton-Villiers-Triumph (NVT). However, this restructuring would result in a number of closures and redundancies. Without warning, in September 1973 NVT Group chairman Denis Poore announced the closure of Meriden works effective February, 1974. Of 4,500 employees, 3,000 were made redundant. Faced with unemployment and having their products handed over to a rival firm, the workers at the Meriden factory demonstrated against a move to Small Heath, Birmingham, the BSA site and staged a sit in for two years. The Bonneville engine size was increased to 724 cc in 1973, and called a 750. Edward Turner died at home in his sleep on August 15, 1973.

Cooperative

As scheduled, Trident production moved to the BSA factory in Small Heath in 1974, but as BSA used non-craft labour in manufacturing, quality fell dramatically. In October 1974 the Labour Government announced the formation of the Meriden Cooperative under Tony Benn, with a loan of £5 million - on the condition that NVT retained ownership of the name, and continued the sales and marketing of the machines. The cooperative resumed production in March 1975, but dropped production of the lightweight T120, to concentrate on the 750 cc twin machines, the Bonneville and the Tiger, primarily for the USA market. The cooperative needed additional cash, and agreed a deal with Lord Weinstock's GEC company to sell 2,000 Bonnevilles for £1,000,000 together with consultation on setting up a sales force.

Meanwhile, NVT stopped production of the Trident in 1975, and also killed off the development of the 1000 cc Quadrent (often and mistakenly called the "Quadrant") due to cash

1971 Triumph Daytona

flow difficulties. A number of key engineers left the company, including Henry Vale, Jack Wickes, Les Williams, Ivor Davies, Arthur Jakeman and Norman Hyde

In 1977, after fighting over who had rights to sell Triumph motorcycles for many years, NVT went bankrupt and the rights were sold to the Meriden Cooperative. The limited edition Silver Jubilee T140V was made to commemorate Queen Elizabeth's 25 years on the throne, a T140 Bonnie with hand-striped wheel rim, chromed engine cases and special sidecover badges. Nominally 1,000 were scheduled for the UK, 1,000 for the US, and about 400 more made for export later. The model sold well, and production increased slowly to 350 machines a week, 60% going to the USA. After this it was all downhill, with no investment in new machines, merely makeovers of the 750 cc twin.

However, the Bonneville T140D won the "Machine of the Year" award in Motor Cycle News – a questionable honour this late in the bike's life, owing more to the bike's reputation than its competency against the (mostly Japanese) competition. The T140D had Lester cast alloy wheels, a new cylinder head with parallel intake tracts, Amal MKII carburettors, Lucas Rita electronic ignition system, and a lower 7.9:1 compression to reduce vibration. In 1980, debt reached £2 million pounds - additionally above the earlier £5 million loan. In October, the British government wrote off £8.4 million pounds owed by Triumph, but still left the company owing £2 million to Britain's Export Credit Guarantee Dept. Triumph experimented with several designs in its last years, none able to stop the decline.

In 1981 the T140D Bonneville Royal Wedding to celebrate marriage of Prince Charles and Princess Diana reached the sales rooms, with 250 each for the UK and America. It had electric start, chrome fuel tank and wheels, and a certificate – and after the original SpeedTwin, the launch Bonneville of 1959 and late 1960s derivatives, is one of the most prized models for a collector.

1982 was the last year of "full" production, with the 8-valve TSS model launched – although a porous cylinder head made by external contractors was its death knell. The company with no money briefly looked at buying the bankrupt Hesketh Motorcycles, and even badged one as a marketing trial – but went bankrupt itself in late 1983.

In 1983 Triumph went into receivership. John Bloor, a 53-year-old plasterer turned wealthy English property developer and builder, who had little interest in motorcycles, had for some time wanted to start up a manufacturing business. Bloor became interested in Triumph, and particularly its still highly regarded brand name. Bloor bought the name and manufacturing rights from the Official Receiver. Enfield India lost, bidding £55,000 pounds to the Official Receiver. A new company Triumph Motorcycles Ltd (initially Bonneville Coventry Ltd), was formed.

Harris Triumph

Because the company's manufacturing plant and its designs were not able to compete against the now-dominant Japanese makers, Bloor decided against relaunching Triumph immediately. Initially, production of the old Bonneville was continued under licence by Les Harris of Racing Spares, in Newton Abbot, Devon, to bridge the gap between the end of the old company, and the birth of the new company. For five years from 1983, about 14 were built a week in peak production - excluding the USA, where due to problems with liability insurance, the Harris Bonnevilles were never imported.

Hinckley Triumph

Bloor set to work assembling the new Triumph, hiring several of the group's former designers to begin work on new models. Bloor took his team to Japan on a tour of its competitors' facilities and became determined to adopt Japanese manufacturing techniques and especially new-generation computer controlled machinery. In 1985, Triumph purchased a first set of equipment to begin working, in secret, on its new prototype models. By 1987, the company had completed its first engine.

In 1988 Bloor funded the building of a new factory in Hinckley, Leicestershire. Bloor put between £70million and £100million into the company between purchase of the brand and broke even in 2000.

A range of thoroughly modern machines using famous model names from the past arrived in 1991. Brand new 750 cc and 900 cc triples and 1000 cc and 1200 cc fours all using a modular design to keep production costs low - an idea originally put forward, in air-cooled form, in the early 1970s by Bert Hopwood but not implemented by the then BSA-Triumph company - were built and proved successful. As sales built, big fours were phased out of the lineup - Triumph's heritage is tied to parallel twins and triples, and these are the marketing and development focus of Triumph's marketing strategy today. Four-cylinder models found themselves competing head-on against Japanese machines, especially in the sports bike market, and although competent could not generate sufficient profit for a relatively low-volume manufacturer like Triumph. In addition to modern machines, Triumph is now also carving out a niche in the motorcycle market based on nostalgic looking engine technologies and design. The 865 cc iterations of the Bonneville and Thruxton look like slightly revised versions of their 1960s counterparts - so although looking and sounding original, internally they include modern valves and counter balance shafts. For their contemporary range of motorcycles, the distinctive triple is Hinckley Triumph's trademark, filling a niche between European and American twins and four cylinder Japanese machinery. The 2294 cc triple Rocket III cruiser was introduced in 2004 and proved highly successful.

In February 2002, as the company was preparing to celebrate its 100th anniversary as a motorcycle maker, its main factory was hit by fire, destroying most of its manufacturing capacity. Nevertheless, the company, which by then numbered more than 300 employees, quickly rebuilt the facility and returned to production by September of that year. Furthermore, in 2003, Triumph opened a new, cutting-edge manufacturing facility in Thailand. Also, assembly and painting facility in Thailand was opened in 2006 by Prince Andrew. Triumph is building another facility in Thailand supposedly to be an engine manufacturing site.

The Triumph Group announced sales of 37,400 units in the financial year ending 30 June 2006. This represented a growth of 18% over the 31,600 units produced in 2005. Company turnover rose 13% to £200 million ($370 million), but net profit remained static at around £10.3 million due to recent investment in production facilities.

TRIUMPH BONNEVILLE

Triumph Bonneville is the name given to three distinct model lines of this notable British motorcycle. They share a parallel-twin four-stroke engine configuration, but the latest motorcycle to carry this name is of a totally new design and is manufactured by the modern successor of the original Triumph company.

The early 650 cc capacity production Triumph T120 Bonneville, often known as the duplex frame model, was replaced in the early 1970s by the T140 Bonneville which was the same basic machine but with a 750 cc engine. Later T120 Bonnevilles used a new frame which held the engine oil instead of a separate tank; this development of the Bonneville became known as the oil in frame version.

The current model Triumph Bonneville uses a modern engine and is unrelated to the original T120 and T140 in design. Since the arrival of this motorcycle, the earlier T120 and T140 have been popularly named 'Meriden Bonnevilles' and the modern version is known as a 'Hinckley Bonneville', reflecting the location of the two factories and to differentiate between the distinct types.

The original Triumph Bonneville was named after the Bonneville Salt Flats in the state of Utah, USA. Although later enlarged to 750 cc, in the late 1970s and early 1980s it suffered when compared to more modern and reliable Japanese motorbikes from Honda and other manufacturers.

TRIUMPH BONNEVILLE	
PRODUCTION	1959 - present
MAX SPEED	-
ENGINE TYPE	Parallel twin
BORE/STROKE	90.0 x 68.0mm
CAPACITY	865cc
COMPRESSION RATIO	-
TRANSMISSION	5 speed
FINAL DRIVE	X ring chain
VALVE TRAIN	DOHC
CARBURETION	Fuel Injection
FRONT TIRE	100/90ZR-19
REAR TIRE	130/80ZR-17
DRY WEIGHT	205kg
WHEELBASE	1,500 mm
SEAT HEIGHT	775 mm

TRIUMPH DAYTONA 675

Triumph Daytona 675 development started in 2000 following the launch of the TT600. The TT600 represented Triumph's first modern foray into the middle weight sports motorcycle market. A decision was made to manufacture a machine closer aligned with traditional Triumph values. A notable technical decision was the selection of a three cylinder engine as the power plant, instead of the four cylinder used by the TT600 and the other 600 cc supersport motorcycles.

In 2001, soon after the completion of the similarly three cylinder powered Triumph Daytona 955i, Triumph began engineering analysis to work out weight, engine performance in horsepower and torque. Pleased with the figures, the project moved to the full concept phase in March 2002.

Initial chassis development work was done using a chopped Daytona 600 chassis. Triumph moved the wheelbase, adjusted the head angle, and modified the tank. This new configuration exhibited better performance than the original Daytona 600, forming a basis to compare against competitive bikes such as the Kawasaki Ninja ZX-6R and Honda CBR600RR. While engine development had not been completed, computer aided chassis development continued with the data collected from these tests.

Design work for the Daytona 675 proceeded, producing a primarily black design based on the Daytona 600. However, this initial design was discarded as great British designs of the 1960's had "a flowing curved design - no sharp angular aggressive edges". A member of the engineering team produced a concept drawing of the 675 as a naked bike. Styling was based upon this concept drawing and that of the earlier T595 model. Styling development continued in house, staying close to spirit of earlier Triumph design. Market research groups made up of a variety of different classes of sportbike riders choose the latter design of bike which was refined and adopted for production.

TRIUMPH DAYTONA 675

PRODUCTION	2006 - present
MAX SPEED	-
ENGINE TYPE	In-line, 3 cylinder
BORE/STROKE	74.0 x 52.3mm
CAPACITY	675cc
COMPRESSION RATIO	-
TRANSMISSION	6 speed
FINAL DRIVE	0 ring chain
VALVE TRAIN	DOHC; 12 valve
CARBURETION	Fuel Injection
FRONT TIRE	120/70ZR-17
REAR TIRE	180/55ZR-17
DRY WEIGHT	165kg
WHEELBASE	1,395 mm
SEAT HEIGHT	825 mm

TRIUMPH SPEED TRIPLE

The new bike was first released to the public in 1994, and in a nod to the historic Triumph Speed Twin was dubbed the "Speed Triple". The original 1938 Speed Twin was powered by a 498 cc vertical twin cylinder engine, and was considered a high performance machine in its day. The new Speed Triple was based on the new Triumph Triple series of modular engines, which also powered the standard Trident, Daytona sportbike, and the Thunderbird retro bike. This engine came in two displacements as a triple; 750 cc for some European markets, and 885 cc for all other markets. The Speed Triple originally was equipped only with the 885 cc engine, but just before significant changes to the bike were made in 1997 a very few 750 machines were produced using leftover Euro spec engines.

Early Speed Triples were all carbureted, and were designated T300 series bikes (Technically, T309). 1994/1995 models came with the standard 885 cc water cooled engine and a rugged five speed transmission. Subsequent Speed Triples all had the same engine with six speed transmissions, except for the brief run of 750 cc bikes. As with all the modular Triumphs, the T309 series Speed Triple had a very large single steel tube backbone frame, and used the engine as a stressed member. Front and rear suspension were fully adjustable, and were made by the well known company Showa. At the rear was a single monoshock with a progressive linkage, and at the front were standard hydraulic forks fitted with dual disk brakes. Colour choices were limited to a basic black, the yellow used on the Daytona, or the iconic Fireball Orange.

TRIUMPH SPEED TRIPLE

PRODUCTION	1994 - present
MAX SPEED	-
ENGINE TYPE	In-line 3 cylinder
BORE/STROKE	79.0 x 71.4mm
CAPACITY	1050cc
COMPRESSION RATIO	-
TRANSMISSION	6 speed
FINAL DRIVE	X ring chain
VALVE TRAIN	DOHC; 12 valve
CARBURETION	Fuel Injection
FRONT TIRE	120/70ZR-17
REAR TIRE	180/55ZR-17
DRY WEIGHT	189kg
WHEELBASE	1,429 mm
SEAT HEIGHT	815 mm

YAMAHA

YAMAHA

Yamaha Corporation

Yamaha was founded in 1887 as a piano and reed organ manufacturer by Torakusu Yamaha as Nippon Gakki Company, Limited in Hamamatsu, Shizuoka prefecture, and was incorporated on October 12, 1897. The company's origins as a musical instrument manufacturer is still reflected today in the group's logo — a trio of interlocking tuning forks.

After World War II, company president Genichi Kawakami repurposed the remains of the company's war-time production machinery and the company's expertise in metallurgical technologies to the manufacture of motorcycles. The YA-1 (aka Akatombo, the "Red Dragonfly"), of which 125 were built in the first year of production (1954), was named in honor of the founder. It was a 125cc, single cylinder, two-stroke, streetbike patterned after the German DKW RT125 (which the British munitions firm, BSA, had also copied in the post-war era and manufactured as the Bantam and Harley-Davidson as the Hummer). In 1955, the success of the YA-1 resulted in the founding of the Yamaha Motor Co., Ltd.

Yamaha has grown to become the world's largest manufacturer of musical instruments (including Pianos,"silent" pianos, drums, guitars, woodwinds, violins, violas, celli, and vibraphones), as well as a leading manufacturer of semiconductors, Audio/Visual, computer related products, sporting goods, home appliances and furniture, specialty metals, machine tools, and Industrial robots.

In October 1987, on its 100th anniversary, the name was changed to The Yamaha Corporation.

In 1989, Yamaha shipped the world's first CD recorder. Since then, Yamaha has purchased Sequential Circuits in 1988 and bought a significant share (51%) of competitor Korg in 1989–1993.

In 2002, Yamaha closed down its archery product business, that was started in 1959. Six archers in five different Olympic Games won gold medals using their products.

It acquired German Audio Software manufacturers Steinberg in 2004, from Pinnacle Systems.

In July, 2007, Yamaha bought out the minority shareholding of the Kemble family in Yamaha-Kemble Music (UK) Ltd, Yamaha's UK import and musical instrument and professional audio equipment sales arm, the company being renamed Yamaha Music U.K. Ltd in autumn 2007. Kemble & Co. Ltd, the UK piano sales & manufacturing arm was unaffected.

On December 20, 2007, Yamaha made an agreement with the Austrian Bank BAWAG P.S.K. Group BAWAG to purchase all the shares of Bösendorfer, intended to take place in early 2008. Yamaha intends to continue manufacturing at the Bösendorfer facilities in Austria.The acquisition of Bösendorfer was announced after the NAMM Show in Los Angeles, on 28 January 2008. As of February 1, 2008, Bösendorfer Klavierfabrik GmbH operates as a subsidiary of Yamaha Corp.

Yamaha Corporation is also widely known for their music program that began in the 1980s.

Yamaha Motor Company

Yamaha Motor Company Limited, a Japanese motorized vehicle-producing company (whose HQ is at 2500 Shingai, Iwata, Shizuoka), is part of the Yamaha Corporation. After expanding Yamaha Corporation into the world's biggest piano maker, then Yamaha CEO Genichi Kawakami took Yamaha into the field of motorized vehicles on July 1, 1955. The company's intensive research into metal alloys for use in acoustic pianos had given Yamaha wide knowledge of the making of lightweight, yet sturdy and reliable metal constructions. This knowledge was easily applied to the making of metal frames and motor parts for motorcycles. Yamaha Motor is the world's second largest producer of motorcycles. It also produces many other motorized vehicles such as all-terrain vehicles, boats, snowmobiles, outboard motors, and personal watercraft.

In 2000, Toyota and Yamaha Corporation made a capital alliance where Toyota paid Yamaha Corporation 10.5 billion yen for a 5 per cent share in Yamaha Motor Company while Yamaha and Yamaha Motor each bought 500,000 shares of Toyota stock in return.

Yamaha has a long racing heritage where it has had its machines and team win many different competitions in many different areas, for example both road and off road racing, also Yamaha has had great success with riders such as Bob Hannah, Heikki Mikkola, Kenny Roberts, Chad Reed, Stefan Merriman and the latest, Valentino Rossi. Yamaha is known to those who are older in age as the designer of the modern motocross bike, as they were the first to build a production mono-shock motocross bike (1975 for 250 and 400, 1976 for 125) and one of the first to have a water-cooled motocross production bike (1981, but 1977 in works bikes).

Since 1962 ,Yamaha produced production road racing grand prix motorcycles that any licensed road racer could purchase. In 1970, Non-factory "privateer" teams dominated the 250cc World Championship with Great Britain's Rodney Gould winning the title on a Yamaha TD2.

Yamaha has made an extensive number of two- and four-stroke scooters, on-road and off-road motorcycles. The Yamaha XS 650, introduced in 1970, was such an overwhelming success that it crippled the British monopoly of vertical twin motorcycles.

YAMAHA YZF-R1

Starting with the FZR1000 in 1987, Yamaha introduced several motorcycle engineering firsts including the boxed aluminum "DeltaBox" frame, advanced intake and exhausts technologies including a 5 valve-per-cylinder head and an exhaust EXUP power-valve. The engines usable power output was dramatically improved throughout the entire RPM range and featured one of the flattest power curves ever seen on a motorcycle. The aluminum Deltabox frame was very light and rigid when compared to steel and both handling and braking were vastly improved over the old-style steel cradle frames. These numerous advances made the FZR1000 a veritable "tour de force" and it was voted "bike of the decade" by many who found it to be quite capable both on and off the track.

For the next four years Yamaha enjoyed significant sales and racing success, however in 1992 Honda introduced the CBR900RR Fireblade which was essentially a marriage of the chassis used for their 600cc motorcycles with an over-bored 750cc engine. Even though the Fireblades weren't as powerful as the FZR1000, it was lighter and shorter which resulted in much quicker handling. It took Yamaha four years to realize the significant weight and power changes that were introduced in the short-lived YZF1000R "ThunderAce". Still, the YZF1000R was based on the original Genesis engine which was canted forward sharply causing the wheelbase to be longer than the Fireblades.

An all-new YZF-R1 for the 2007 model year was announced on 9 October, 2006. Key features include an all-new inline four-cylinder engine; going back to a more conventional 4-valve per cylinder rather than Yamaha's trade mark 5-valve genesis layout. Other new features are the Yamaha Chip Control Intake (YCC-I) electronic variable-length intake funnel system, Yamaha Chip Control Throttle (YCC-T) fly-by-wire throttle system, slipper-type clutch, all-new aluminum Deltabox frame and swingarm, six-piston radial-mount front brake calipers with 310 mm discs, a wider radiator, and M1 styling on the new large ram-air ports in the front fairing. 2008 brought BNG and the ability to buy limited edition Fiat plastics.

YAMAHA YZF-R1

PRODUCTION	1998 - present
MAX SPEED	-
ENGINE TYPE	In-line 4 cylinder
BORE/STROKE	77 x 53.6mm
CAPACITY	998cc
COMPRESSION RATIO	12.7:1
TRANSMISSION	6 speed
FINAL DRIVE	O ring chain
VALVE TRAIN	DOHC, 16 valve
CARBURETION	Fuel Injection
FRONT TIRE	120/70-ZR17
REAR TIRE	190/50-ZR17
DRY WEIGHT	176kg
WHEELBASE	1,414 mm
SEAT HEIGHT	835 mm

YAMAHA FZ1

2001 was the first year that the FZ1 was produced. Models produced in the period 2001 to 2005 were known as FZS1000S (Fazer in Europe). They had a modified Yamaha YZF-R1 motor in a steel tubular frame. The FZ1 was carburated and produced around 140 horsepower. They were virtually unchanged over this period, except for different color options. In some European countries the 2005 model saw the introduction of an exhaust based catalytic converter, albeit of a rudimentary design.

2006 saw the introduction of a completely new model. The main changes included a new chassis, suspension, body work and a completely new engine, never seen before in the big Fazer. This brought the bike up to date with modern rivals. There have been instances of fuel injection glitches on the new model, although there are various 'fixes' available. The 2007 and on models have resolved most of these early fuel injection problems. The 2006 model has a 998 cc DOHC 20-valve R1 engine, which produces 150 horsepower at 11,000 rpm, set in an all-new compact aluminum frame.

YAMAHA FZ1

PRODUCTION	2001 - present
MAX SPEED	-
ENGINE TYPE	In-line 4 cylinder
BORE/STROKE	77 x 53.6mm
CAPACITY	998cc
COMPRESSION RATIO	-
TRANSMISSION	6 speed
FINAL DRIVE	0 ring chain
VALVE TRAIN	DOHC; 20 valve
CARBURETION	Fuel Injection
FRONT TIRE	120/70ZR-17
REAR TIRE	190/50ZR-17
DRY WEIGHT	199kg
WHEELBASE	1,460 mm
SEAT HEIGHT	815 mm

YAMAHA FJR1300

The FJR1300 was introduced to Europe in 2001 before arriving in North America in 2002 with the 2003 model year designation and offered in a non-ABS version only. Motorcyclist magazine named the 2003 model Motorcycle of the Year. It had 298 mm front rotors. It appeared in Europe in various colours: silver, blue, black and red.

The 2004 European model came in a range of colours, including Silver Storm. The 2004 North American models included both a non-ABS version with traditional blue anodized brake calipers and a new ABS version. Both are Cerulean Silver. Other refinements included an upgrade to the suspension rates, 320 mm front brake rotors, and a fairing pocket for small items.

The 2005 North American model year remained structurally unchanged with a non-ABS and ABS model in Galaxy Blue.

In 2006 the U.S. and World model years synchronized and design significantly changed including trailing arm changes, radiator curving, instrumentation changes, upgraded alternator and significant attention to airflow changes from reported heat issues in previous years. In response to these complaints, Yamaha added several adjustable vents to the FJR1300, allowing the rider to direct air to or away from the body.

The base FJR1300A model has ABS and is Cobalt Blue while the Cerulean Silver colored FJR1300AE model features a semi-automatic transmission. The AE model has YCCS, or Yamaha Chip-Controlled Shift. The rider can either utilize the standard foot shifter sans a clutch lever, or shift via a lever on the left bar where a clutch would normally be. The AE model continues in production through 2008.

For 2008 some minor changes were introduced, including an update to the altitude-related ECU issues and throttle 'feel', notably to improve low speed on/off throttle transitions. The colours announced in Europe are; Silver (Silver Tech), Black (Midnight Black) and Graphite. 2008 also sees minor changes in the ABS system. Four FJR1300s finished in the top ten of the 2007 Iron Butt Rally.

YAMAHA FJR1300

PRODUCTION	2001 - present
MAX SPEED	-
ENGINE TYPE	In-line 4 cylinder
BORE/STROKE	79 x 66.2mm
CAPACITY	1298cc
COMPRESSION RATIO	10.8:1
TRANSMISSION	6 speed
FINAL DRIVE	Shaft
VALVE TRAIN	DOHC; 16 valve
CARBURETION	Fuel Injection
FRONT TIRE	120/70ZR-17
REAR TIRE	180/55ZR-17
DRY WEIGHT	264kg
WHEELBASE	1,539 mm
SEAT HEIGHT	800-820 mm

YAMAHA V STAR 1100 CLASSIC

The Yamaha DragStar 1100, sold as the V-Star 1100 in North America, is a motorcycle manufactured by Yamaha Motor Corporation. It comes in two versions, the XVS1100 and the XVS1100A, the former a more modern style, and the latter a more classic style, with rounder edges. The seat height is slightly lower on the custom.

The DragStar began as the XVS650 in 1998, and grew in 1999 to include the XVS1100 Custom. The 1100 used a reworked version of the venerable Virago 75-degree, air-cooled v-twin which had been in use since the early 1980s. The Star version offered better torque for the new midsize cruiser bike. The Star carried over the shaft-drive layout from the Virago, but relied on a new suspension and frame, discarding the outboard dual shocks and stressed-member engine arrangement of the Virago in favor of a single-shock and twin downtubes. A single Keihin 37 mm carbuerator with throttle-position sensors is used to efficiently meter fuel to the 8.3:1, oversquare engine.

The DragStar/V-Star gathered a huge following early on, which led to the addition in 2000 of the Classic model. As opposed to the Custom's bobbed rear fender, exposed forks, 110 front tire and other custom touches, the Classic had longer, more voluptuous fenders, floorboards, beefy fork covers and a 130 front tire. Later a Silverado model, which included such amenities as a windscreen, sissy bar, soft sidebags and the like, was introduced as an upscale model.

YAMAHA V STAR 1100 CLASSIC	
PRODUCTION	1999 - present
MAX SPEED	-
ENGINE TYPE	V twin
BORE/STROKE	95 x 75mm
CAPACITY	1063cc
COMPRESSION RATIO	8.3:1
TRANSMISSION	5 speed
FINAL DRIVE	Shaft
VALVE TRAIN	SOHC; 2 valve
CARBURETION	Fuel Injection
FRONT TIRE	130/90-16
REAR TIRE	170/80-15
DRY WEIGHT	272kg
WHEELBASE	1,645 mm
SEAT HEIGHT	708 mm

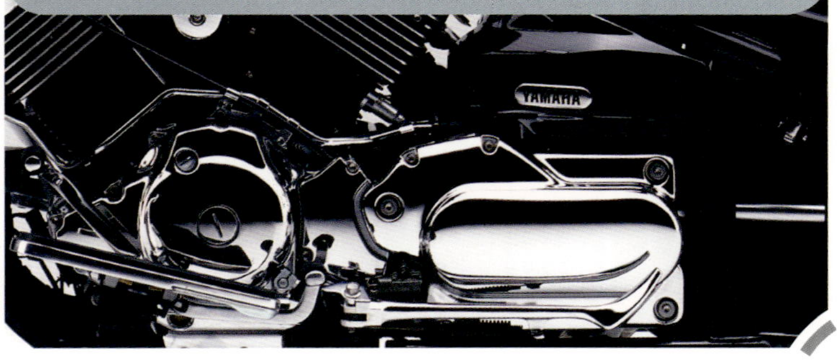

INDEX